Harry Kirke Swann

Nature in Acadie

Harry Kirke Swann

Nature in Acadie

ISBN/EAN: 9783337026493

Printed in Europe, USA, Canada, Australia, Japan

Cover: Foto ©berggeist007 / pixelio.de

More available books at **www.hansebooks.com**

MELVILLE ISLAND.

BY

H. K. SWANN

Author of " The Birds of London," &c.

London

JOHN BALE & SONS

OXFORD HOUSE

GREAT TITCHFIELD STREET, OXFORD STREET, W.

1895

PREFACE.

LITTLE apology is necessary for the appearance of this book ; inasmuch as it is not intended to be in any sense a scientific work. It is more strictly the simple narrative of a Nature-lover's first voyage westward, together with some few attempts at the word-picturing of what he saw during his sojourn in Acadia, which in prose is rendered Nova Scotia—a sweet name, maybe, but not so sweet as the elder one.

Neither is the book intended to be concerned with the doings of men, and it will be found, indeed, that the author has studiously ignored the subject. One need not journey three thousand miles to study *human* Nature.

It will be found, however, that some little attention has been paid to ornithology, and in this connection it is probable that the work may contain something of interest to the naturalist as well as the Nature-lover.

H. K. S.

H. D. THOREAU.

MANY long years have fled
  Since thou wast gone—
Gone! aye, yet still not dead;
  Thou livest on!

In every zephyr breeze
  That wanders lone,
Whispering among the trees
  With listless tone—

Striking on harpstrings free
  Sweet sylvan chords,
While every list'ning tree
  Breathless applauds—

In all the songs of birds,
  Mid woodlands lone,
I hear thy noble words
  Sadder of tone.

E'er through their music throng,
  It seems to me,
Whispers of heavenly song
  That speak of thee.

And in the rippling streams
  That softly sing,
Thy voice for ever seems
  Through them to ring.

＊　　　＊　　　＊　　　＊　　　＊

See where the winding creek
  Pierces the land
In a clear silver streak,
  Woods on each hand!

There as the wavelets swell
  At break of day,
A message strange they tell
  Idly at play—

Burthen so free and far,
  So deep and wide,
Lisped to the truant star
  Caught by the tide—

Message of Spring and Youth,
  Decay and Death ;
Echoes of secret truth
  That the gay south wind saith
  Under its breath.

And e'en the birches slight
  Along the shore,
That from the waters bright
  Their secrets store,

Speak their young modest mind
  With whispers soft,
That an eavesdropping wind
  Carries aloft
  Unto me oft—

Bids me to list the song,
  Sung low and faint
By winds that sigh along
  Sad in complaint ;

Plaint of a sorrow rare
  In whispered tone ;
Murmurs and sighings their
  Meaning unknown ;
Babbled by leaves and air
  And brooklets lone.

                              H. K. S.

NATURE IN ACADIE.

CHAPTER I.

Have the elder races halted?
Do they droop and end their lesson, wearied over there beyond the seas?
We take up the task eternal, and the burden and the lesson,
Pioneers! O pioneers!—*Whitman.*

H for a larger field—the solitude of the forest, or the silence of the prairie, where one can roam at will and lose the rush of life! Such was the hope that found me on the deck of an outward bound liner in the docks at Liverpool on this 29th day of September, buoyed with the hope of the newer world across the Atlantic. Long years had I thirsted for the vast solitary woods, the clear sparkling lakes and silent hills of that great new land, and now at last came the moment of departure.

Clamour and confusion reign supreme. The deck of the great steamer is crowded with passengers taking a last view of the old country, which they are leaving, perhaps, for many years, possibly for ever! Before we are clear of the docks night succeeds to the calm glory

of the autumn evening, and when we are fairly out in
the Mersey the upper deck rapidly becomes deserted.
Gradually the town is left behind, and the rows of lights
on either shore grow fewer and fainter, giving place to
night, still and dark, broken only by the screech of an
occasional steamer and the regular thud of the engines.
The stars are now peering down upon the murky black-
ness of the river ; the sea-breeze freshens, and I retire
below, soon to be lulled to sleep by the gentle rolling of
the vessel and the regular " swish " of the waves against
her sides.

The next morning I was on deck soon after daybreak,
the weather being fine, but rather hazy, and the sea just
a trifle " choppy." We were passing through the Irish
Sea the greater part of the day, but although I was
on deck until night I observed hardly any birds; there
being but a few gannets and a small number of the
curious Manx shearwaters, while once a lesser black-
backed gull passed us.

We arrived at Queenstown in the small hours of the
morning, leaving again before daybreak, and until about
mid-day the south-west coast of Ireland was in view at
no great distance. The coast here was one continuous
line of undulating hills with precipitous cliffs, without a
tree visible, but clothed everywhere in a vegetable
carpet of beautiful and varied hues, while here and
there tall rugged rocks rose sheer up from the glassy
surface of the ocean, and in the dips of the coast a few
cabins could be perceived, appearing like tiny white
specks on the hillsides. The weather all day was fine
and bright and the pure sea air peculiarly exhilarating.

I did not notice many birds, there being but a few
gannets, or solan geese, shearwaters and guillemots,
with one or two of the small black petrels. In the
afternoon, however, when we had fairly lost sight of
land, a small bird flew over the vessel and seemed
about to settle, but then darted away. It appeared to
be some species of *Anthus* or pipit.

After leaving the Irish coast things changed for the
worse, and for the next six or seven days we were

labouring against terrible seas which washed the deck from end to end and tossed the great vessel about like a cork in their fury. However, after some days the weather, although still boisterous, quieted down sufficiently to allow me to remain on deck, and I was thus enabled to continue my observations.

The only birds which I observed in the mid-Atlantic were the fulmars, a considerable number of which were following the vessel, having been increasing in numbers daily since we left the Irish coast. They followed persistently in the wake of the vessel, wheeling and gliding with placid flight hour after hour, and every now and then dropping down upon the refuse which was floating in our track. They frequently settled on the water, floating buoyantly upon the waves, and now and then paddling vigorously towards some floating bit of offal, and on seizing it being pursued through the water by several other individuals who endeavoured to tear the prize from the fortunate possessor. They very often flew close alongside, but I never heard them utter any kind of note; most of them had the head, neck and under parts pure white, but in one or two they were of a greyish tint, there being light and dark phases of this species.

The day before reaching Newfoundland was much finer; the vessel moved steadily through the water, making good headway, and the sun shone quite warmly. We were passing over the famous Newfoundland Banks, and the quantity of bird life was surprising, affording undeniable signs of the proximity of land and also of the abundance of congenial food.

Large numbers of kittiwakes were accompanying us; there appeared to be quite 200 of these pretty little gulls—the majority being in immature dress—and they first made their appearance three or four days before, when I noticed four or five of them among the numbers of fulmars which then accompanied us; since then the present species had increased in numbers, while the fulmars had decreased, and I observed but one or two of the latter on this day. The flight of these little gulls is

light and graceful in the extreme, they usually fly close
to the water, upon which they frequently settle, but they
often rise higher in the air and wheel over the vessel, or
several at a time come close alongside, flying close to
the surface of the water, their sharply curved wings
moving with a quick regular beat ; usually, however,
they are to be seen stretching in an irregular column
far away in the vessel's wake like white specks, alter-
nately rising and falling, sweeping onward or settling in
the long line of foam which marks her track.

I also observed here several small parties of the little
auk or "dovekie"; these little birds fly heavily, close
to the surface of the water, and in a compact body;
they are easily recognised by their black and white
plumage, and are usually seen swimming close together
upon the surface of the water, but they dive instantly
upon the slightest alarm, and remain below a considerable time, reappearing at some distance.

About noon a small bird flew on board in an exhausted
condition and settled on the rail of the bridge, afterwards flying down to the deck, where it picked up a few
crumbs, but on my approach it flew away again and
was seen no more. I recognised it as a snow-bunting
in summer dress, but the appearance of this little wan-
derer at such a distance from land was a matter of some
surprise to me. I noticed in addition one or two guillemots and petrels, and also several jaegers or skuas,
including the long-tailed or Buffon's skua, the flight of
which is swift and graceful. The porpoises, of which
I observed small parties every day until now, had all
disappeared.

The morning of Sunday, October 11, broke in a day
of singular fineness, and found us lying at rest in the
land-locked harbour of the quaint little town of St.
John's, Newfoundland. I lost no time in getting on
shore, and immediately started on a tour of investigation
in the surrounding country. The day was cool and
bright, but far warmer and drier than I had expected,
and the sun sailed high in a cloudless blue sky.

Once clear of the little town I directed my steps

towards the woods lying in its rear, and found everywhere much of interest.   The land here is mostly under
cultivation, but it is very poor, the most noticeable pro-
duction being stones, which cover the fields and are
heaped up on the roadsides, while all the walls and
boundaries are made of slabs and pieces of stone piled
one upon the other, and all the farmhouses and buildings
are either made of wood or stone.   The stone is mostly
sedimentary deposits of various kinds.

There were very few birds about here, but I noticed a
sparrow in a little plantation which may have been the
chipping sparrow so well known in the States.   I observed several small parties of the American robin or
migratory thrush ; they flew high and were rather wary,
settling in the thick fir plantations where they appeared
to feed.   I also noticed here a few American crows
flying overhead.   Walking back in the direction of the
sea I came on a small lake by the side of which I disturbed a spotted sandpiper, which allowed a very near
approach before taking to flight, when it darted off
uttering a shrill *peet, peet*, and settled again at the water's
edge at some distance.   The day had now turned out
cloudy, but so mild was it that I caught several small
moths here, including a " vapourer," exactly resembling
the species (*O. antiqua*) which is to be seen so commonly
in the streets of London.   Up in the hills near by I disturbed a single fox sparrow, a large handsome species,
having the lower part of the back and the tail of a
reddish colour.

The next day I made my way on to the hills stretching
away along the coast to the southward.   There is a
very extensive tract of rocky and mountainous land here,
which presents much resemblance to the Highlands of
Scotland, both as regards its flora and its picturesque
appearance—hills and dales, covered with great boulders
and protruding rocks, alternating with peaty bogs, stag-
nant swamps or clear quiet mountain lakes, with here
and there a rushing stream or miniature waterfall ; and
this extending as far as the eye can reach, the only sign
of the proximity of man being yon hillside on which the

pine stumps stand out grim and hoary, marking where
the destroying axe has been at work. The forests of
the island appear to be chiefly coniferous, and on the
coast the trees do not attain to any considerable size,
although there is a luxuriant undergrowth. In the
great unexplored interior, however, are many tracts of
forest and also some very extensive lakes.

A melancholy history attaches to the former natives
of Newfoundland, for of the once numerous and powerful race of aborigines throughout the length and breadth
of this great country, not one remains. The colony of
Indians on the west coast of the island belong to the
Mic-Mac tribe, to whose persecution, added to that of
the dreaded paleface, the extinction of the Beoths
was due. Early in the present century proclamations
protecting the Beoths, as these aborigines called
themselves, were issued by the British Government,
but as usual they came too late, for a very few
years after saw the final extinction of these ill-fated
people. Rumour, indeed, has it, that the last of the
Beoths, a mere handful, passed across the Strait of Belle
Isle in two canoes early in the present century, and
landing on the opposite coast of Labrador, disappeared.

We left St. John's about noon on Tuesday, and during
the remainder of the day were skirting the coast of
Newfoundland in a southerly direction. I observed a
fair number of seabirds of different kinds, including a
number of my recent acquaintances, the kittiwakes, and
also several great skuas.

We passed the mouth of the Gulf of St. Lawrence
during the night of October 14, and all the next morning
were steaming down the coast of Nova Scotia, within
twelve or fifteen miles of land, and with the surface of
the water almost unruffled, so delightful was the day.
The coast presents an almost unbroken line, dipping
here and there into a valley where some little river
enters the sea, and dotted at frequent intervals with
tiny white houses, succeeded perhaps by scarcely distinguishable fields or wooded and sterile hills, with here
and there a church spire rising in the distance. On

such a day one is tempted to forget that this is the " mournful and misty Atlantic," and that yonder coast is the home of those treacherous sea fogs, which are unfortunately far too frequent here.

Presently a dark line was distinguishable right ahead which soon resolved itself into a rocky point stretching out to the southward and known as Chebucto Head. The vessel's course had been altered, and we were now rapidly nearing the entrance to Halifax harbour. The sea here was unruffled and glittered like glass in the rays of the declining autumn sun, while the air was so remarkably clear and still as to produce a curious optical illusion, for the land appeared to be but a comparatively few yards from one, while it was in reality more than a mile away. Soon we were passing by the North-West Arm, winding serpent-like between its sloping and wooded shores, all gorgeous with the varied autumn foliage of the trees, and ere long we were lying alongside the wharf in the harbour of Halifax.

CHAPTER II.

HE harbour of Halifax is said to be, perhaps, the safest and best in the world, and it certainly seems to deserve this eulogy, nature having rendered it impregnable to the elements, and man having done the same as regards his fellow-man. It is five or six miles in length, and is connected at the north end, by a narrow arm, with Bedford Basin, a great forest-lined sheet of water six miles long by four in width, and capable, according to the guide books, of containing all the navies of the world.

Halifax, the capital of Nova Scotia, differs not from the ordinary run of civilised cities, there being the usual proportion of substantially-built brick or stone houses sheltering the rich, and the usual proportion of wooden cabins sheltering the poor. It stretches for some three miles along the western slope of the harbour, and is terminated in the south by the entrance to the North-West Arm, which circles round for some three miles in the rear of the town. This picturesque inlet is about a quarter of a mile in breadth, with abruptly sloping sides, thickly wooded for the most part—except on the town side, where some of the " well-to-do " have pitched their habitations, and converted its less abrupt shores into pastures and gardens. From the head of the Arm a low valley continues the circuit round to an angle at the south-west end of Bedford Basin, at a spot known as Three-Mile House, from the fact of its being situated at a distance of three miles from Halifax.

Near the head of the Arm, and on the side away from the town, another pretty little inlet opens out of it, forming a shallow basin in the midst of wooded heights that tower around it; at the mouth of this inlet is a tiny island, occupied by the military, and known as Melville Island. Dense woods climb the steep sides of this extremely picturesque little basin, obtaining a precarious footing among the giant boulders or fragments of granite which are piled in confusion everywhere, or with their roots uncovered by the fury of one of the numerous torrents which descend these rugged slopes after a rainstorm, tearing huge boulders and blocks of granite from their beds, and launching them with the momentum of an avalanche down their furrowed channels, accompanied by a shower of the *débris* which is for ever finding its way down to the already shallow basin below. It is curious to see these doomed trees, half uncovered by the violence of the flood, yet clinging with all their remaining arms, or roots, to the torn and jagged sides of the watercourse, like a despairing swimmer, grasping with his ebbing strength some jutting rock in the vain hope of escape from the relentless torrent which bears him swiftly away.

On climbing to the ridge above, one finds oneself standing upon a more or less revealed plateau of granite, forming the crest of this mighty upheaval, and running away on either hand in ribs or buttresses, which form miniature valleys between, filled with a dense undergrowth springing up between the scattered granite boulders; or with solemn moss-draped firs, and pines, and hemlocks, staggering up the rocky slopes and standing out triumphantly here and there on the arid crown, alternating now and again with graceful maple and birch, while sometimes one looks down into a veritable Valley of Death, the vegetation having had its brief day and moved on elsewhere, leaving gaunt, lifeless stumps, or prostrate and whitened stems and limbs, for all the world like a great littered mass of bleached bones. But up from this great charnel-house even now are springing new undergrowth and lusty youthful trees and herbage, and soon the roving birds will come here again, and the little red squirrel will make his nest here once more when these firs shall have reared their proud heads over this rampart of granite to look on the decay of these stately neighbours of theirs.

Even on the crest of the hill old motherly Nature has covered up the grim mass of granite with a scanty film of *débris*, and watered it with tears of rain, and the forest has crept slowly up and across, and hidden the naked rock under its ample cloak, except here and there where a rugged patch stares barrenly out from the verdant fold of moss which borders it around and creeps slowly and steadily, year by year, up it from all sides, while the rock's furrowed back is for ever wearing and crumbling down to meet it. These are the haunts of my little favourite, the black snowbird, and whenever you approach softly up to one of these little rock-patches, you are almost sure to see this trim little bird hopping daintily, like our own familiar robin, over the harsh granite, now and then pecking in a half-hearted way at something, or perching motionless, with a seeming gentle contentment that sends an indescribable feeling through you when you involuntarily raise your eyes and glance around at the vast, silent forest—without a path, a habitation, and hardly a sign of animal life—which closes in around you.

All the way along from the heights above the North-West Arm, and above the valley previously mentioned, and above the great Bedford Basin, one vast continuation of granite ridges, and forest, and lake, and scrub-covered upland, runs back for miles upon miles, and here it is that most of my spare time during my stay at Halifax was spent, and most of the information given in the following pages was gathered.

CHAPTER III.

URING a short walk in the vicinity of Halifax, on the day after my arrival, I observed our European house-sparrow to be very common about the town ; in fact it seemed even as impudent and abundant as in the streets of London, they having, no doubt, now been naturalized long enough to look upon these American cities as their rightful inheritance.

I observed numbers of the American crow about the fields in the immediate neighbourhood of the town. This bird very much resembles the European species (*C. corone*), but, to my ear, its note is somewhat different, being sharper and more querulous, and at a little distance having a very great resemblance to the bark of a small dog. By the roadside, among the fields, I disturbed two or three American pipits, a bird which has much of the habits and appearance of our meadow pipit (*A. pratensis*) and, to an English observer, seeming to be almost the same bird. These individuals were evidently only visitors on their southward migration, their breeding range being to the northward, in Labrador and up to the Arctic regions.

On the morning of October 18, a fine, sunny day, I started out for a ramble in the direction of the North-West Arm, which I reached after a walk of nearly three miles. At this season the beautiful and varied hues of the autumnal foliage formed quite a

striking spectacle as I approached the woods, and one
which I do not think I have ever seen equalled by our
English woods.

Passing the head of the Arm, I ascended the opposite
hill by the St. Margaret's Bay road and soon came to
Chocolate Lake, a small-sized lake, not more than a
quarter of a mile in breadth, lying to the left of the
road and shut in by private lands.   The granite begins
to appear at the surface just here, but I noticed that
near the head of the Arm the surface is underlain by a
large amount of dark-looking sedimentary rock, strongly
impregnated with iron, and apparently formed as a deposit in some former estuary or mouth of a river.   It is
in a fragmentary condition and mixed with clayey loam,
giving it a "pudding-like" appearance, and suggesting
the action of some mountain torrent (or, perhaps,
glacier) in the rounded form of the fragments.   Towards Halifax, however, almost the whole hill consists
of this same kind of rock upheaved bodily.

On the outskirts of the forest I observed several
juncos, or black snowbirds, hopping about the roads, but
on my nearer approach they quickly vanished into the
undergrowth.   This interesting little bird is almost
silent except for an occasional slight chirp; its plumage
is sober but pleasing, the bill being yellowish-white,
entire upper plumage and the throat and breast dusky
slate colour, and the abdomen and outer tail feathers
white; in the female the upper plumage is greyish-
brown instead of dusky slate.

Continuing along the road I came upon a series of
beautifully clear and sparkling lakes, hemmed in by the
picturesque forms of the granite hills, clothed with their
rich autumnal garb of many-hued foliage of the scrub
and underwood.   These lakes drain from one to another
for miles, Chocolate Lake being the termination of the
series.   The road here ran for some distance along the
water's side, while on the left rose precipitous wooded
heights with occasionally a patch of low swampy ground
intervening, covered with a dense growth of spongy
moss and filled with swamp-loving bushes.

The day being particularly fine and warm many insects were abroad in spite of the lateness of the season, which, however, hardly corresponds to the chilly and windy October of old England, for here the autumn is the most beautiful and enjoyable time of all the year, and is usually continued, in days of singular fineness, right up to the middle of December.

I found, however, but two species of butterflies still remaining. One of these (*Colias philodice*) much resembled the pale clouded yellow (*C. hyale*) of Europe, and I met with it in some abundance on the roadside near a strip of pasture land. The other species (*Vanessa milberti*) presented a good deal of resemblance to the familiar lesser tortoise-shell (*V. urticæ*) of the old country, but the markings on the basal half of both anterior and posterior wings constituted a clearly-defined patch of black, with only a faint indication of reddish markings on the anterior wings, while the remainder of each wing was unspotted reddish orange, with the edge of the wing similar to that of *V. urticæ*, except that there were no white markings. The specimen which I obtained flew from over the wall of a garden close by and glided along by a bank on the roadside, much as our English variety would do. *Vanessa milberti* appears to be spread over the larger part of North America, being found across the United States to the Rocky Mountains, but it does not appear to be very common in Nova Scotia.

This similarity to the old-world fauna is just as evident in the *Heterocea* or moths. There is, for instance, a species of *Abraxas* common enough about Halifax which can scarcely be distinguished by a casual observer from the "currant moth" (*A. grossulariata*) of England.

I met with a small species of the Libellulidæ in considerable abundance on the roadsides about the forest. The male insect had the abdomen above of a bright ruby colour, while the female had that part reddish brown, the abdomen being slightly depressed and thicker than in the *Agrions* of Europe, while the insect was also larger than the typical examples of that genus.

Another much larger species of "dragonfly" ("devil's darning-needles" they are called here, as in some parts of England) was also not uncommon along the sides of the lakes. It has the characteristics of *Æshna* and is nearly three inches in length and four inches across the extended wings, which latter are perfectly transparent; the slender cylindrical abdomen is blackish with several transverse markings of a pale blue, and with two stripes of yellow upon each side of the thorax. The flight of this insect is remarkably swift, and when struck at with the net and missed it darts off with such velocity that the eye can scarcely follow it. Sometimes it comes gliding down the rocky gorge or sweeping across the lake with a gentle movement of its wings, and then, suddenly darting upward, it snaps up a fly with an audible click of its jaws and glides on again in search of fresh prey.

Among the rocks and boulders on sunny slopes I found several marbled locusts (*Locusta marmorata*). When settled this insect presents a dusky appearance, the upper wings being mottled with dusky brown, but the lower wings are pale yellow, with a darkish outer margin, and are very noticeable in flight. When flying this insect produces a loud and peculiar "clicking" sound; it flies in an irregular manner and usually but a short distance, settling on the bare surface of one of the boulders, with which its mottled appearance when at rest somewhat harmonizes; if disturbed it leaps some distance like a grasshopper. The latter insects also abounded and I obtained specimens of two or three species.

In a little swamp I obtained a wasps' nest of a kind that is somewhat common in these parts. This specimen was at a height of about five feet above the swamp, and was fixed to a spray in the upper part of a bush, the twig passing through one side of the globe, and the small shoots and leaves being also carefully worked into the wall of the nest. It was about five inches in diameter and almost globular, the outer wall being composed of several distinct and separate layers or envelopes of a peculiarly thin, greyish-white kind of wasp-paper; in the lower part was a small circular opening through which the empty comb could be seen in the interior. Others of these nests that I have seen have been in bushes and saplings, all at moderate heights, and in size usually not much exceeding the one above described, although I have seen one fully ten inches in diameter.

On October 19 I paid the first of many visits to the veteran naturalist, Andrew Downs, C.M.Z.S., of Halifax, who was then in his 81st year, although still comparatively well both in mind and body. His long and extensive experience and acquaintance with many of the older naturalists—dating back to Audubon in America, and Waterton, Gould, Jardine and other naturalists of the old school in England—imparted an especial interest to our conversation. To his credit be it said that his collection of birds stands almost unrivalled in point of workmanship and mounting. A prominent group, in a case of Corvidæ, was a pair of ravens with their nest and eggs, taken near the North-West Arm in the vicinity of Halifax, and in this case also are many medals and awards gained at former exhibitions on both sides of the water. Among other living birds were a fine pair of silver pheasants, the female of which is small and soberly clad and altogether unlike the beautiful male; also a purple or Martinique gallinule, which I observed could perch and climb among the slender twigs of a tree with considerable facility, and one of our old homely birds, the blackbird. He also showed me an exceptionally fine pair of snowy owls, the female of which I found measured fully twenty-seven inches in length, and was more heavily marked than the male, which was also appreciably smaller.

*   *   *   *   *

Since my return to England the news of the death of this simple-hearted and kindly old naturalist has reached me, and it is with sincere pleasure that I find myself enabled to inscribe these few lines, in default of better tribute, to the memory of one who will live long in the minds of all those who knew him. He was essentially a working naturalist, and so leaves very little to retain him in the recollections of the younger generations ; but his collection of birds, presented shortly before his death to Dalhousie College Museum, will perhaps form some slight memento of his work.

CHAPTER IV.

HE 21st day of October opened fairly fine, and the Fates seemed forbearing as I made my preparations for a long day in the forest. I was early astir, and the inhabitants seemed to be still deep in their slumbers as I wended my way past the few houses at the head of the North-West Arm.

With the exception of a few crows clamouring afar off as they straggled away in search of an early breakfast, the first living thing that met my gaze was a pretty little " chipping squirrel," " chipmunk," or ground squirrel as it is variously called. There he sat, on the bottom rail of the fence at the roadside, holding a nut or berry in his little fore-paws, with his fearless gaze turned round upon me as if to question the right of my intrusion—but without a trace of fear. All the stories and traditions of this little animal, treasured up for generations in the mighty country that overshadowed me even then, rushed through my mind as I beheld the little " chipmunk " for the first time in his native haunts. How different he and his surroundings were to the captives or the "specimens " I had hitherto only known ! My eye wandering over the scene took in the fearless little creature sitting there before me on the rough-hewn rail, with the straggling bushes on either hand, while behind the grassy strip of meadow sloped down to the dark basin of Chocolate Lake, with the

2

little ripples sparkling gaily in the morning sun, and all around towered the hanging woods of hemlock and pine. The moment I turned to continue on my way again he dropped off the rail and disappeared in an instant.

In the woods I noticed a good many chickadees, a most lively and scolding little black-and-white-headed species, with all the habits of the English titmice; it appears to me to be intermediate between the coal and marsh titmice of England. Sometimes associated with these were a few golden-crowned kinglets (*Regulus satrapa*) which might readily be taken for the European goldcrest (*R. cristatus*), so similar is it in appearance and habits. These little birds were creeping about in the firs and uttering from time to time their feeble note, just as their old world cousins do.* Another representative of one of our familiar birds which I observed here, was the brown creeper, which is almost identical with the European species; it was creeping up the stem of a fir and uttering its shrill, slight note from time to time.

After passing over a zone of granite hillocks, of such extraordinary irregularity that I had to leap and clamber from boulder to boulder to make any progress, I came upon a secluded little bog at the head of a small lake, which was shut in on all sides by rocky and wooded slopes, but connected at the farther end by a small rocky channel with another swampy forest lake. This bog had evidently formed a shallow part of the lake at no very distant period, but it was now almost dry, except at one part, where a small sluggish stream flowed into the lake, springing mysteriously from the bowels of the eternal granite, not a hundred yards away, in the form of limpid crystal, but assuming an inky appearance as it oozed slowly through the treacherous bog. This stream necessitated very gingerly treading on my part

---

* The most important difference between these two species appears to me to be in the greater size of the American bird, which is longer by half an inch than the European, while the bill is also somewhat stouter than in the latter. Like the European species, the female *R. satrapa* has the crest bright yellow, instead of reddish-orange as in the male.

in crossing it, as the apparent *terra firma* on either side was merely quaking peat-moss and swamp vegetation, and trembled beneath my foot, while the water oozed up as though it were a great sponge, and everywhere were little intersections and veins of dark-looking water. I espied a very handsome frog, having the under parts reddish, just here, but before I could catch it, it disappeared in the evil-looking water of the stream.

As I was leaving the bog, a marsh hawk came flying slowly over. Such places as these are the favourite haunts of this bird, as it can there find abundance of prey in the shape of lizards, mice, small birds, or the larger kinds of insects. This species is closely allied to the hen harrier (*Circus cyaneus*) of the old world; its flight is slow and steady, and not very unlike that of its half-cousins, the owls.

There were abundant signs of the presence of " rabbits " (*Lepus americanus*) in all the more open spots here, yet I only saw one of these animals during the day, and that was extremely shy. In England this species is known as the " American hare," but throughout Canada and the United States it is almost invariably known as the "rabbit," and it is in fact not much larger than our rabbit, although possessing the habits and appearance of the hare; it also has the same leaping gait as the latter, the hind legs being particularly long.

From here I passed over the adjoining ridge, and after fording a little stream by means of the slippery, moss-grown boulders projecting from its bed, and forcing my way through the dense swamp bushes on the other side, I found myself on Long Lake, which is a lake of some considerable size and quite the largest one in this neighbourhood.

The spot I had reached was evidently one of the most secluded on the entire lake, solitary as it was, and having got to a drier and more open situation, I paused awhile to view the wild and lonely scenes which surrounded me. From here I could follow the winding of the lake for nearly a couple of miles ahead, where it became lost to view among the wooded hills.

On my right, the arm I was upon was shut in by the
granite ridge I had previously crossed, while to my
left the shore was lower and clothed with dense woods
which descended right to the water's edge, where the
monarchs of the forest struggled with the dwarfish and
mis-shapen underwood for a footing among the dripping
boulders, draped with green moss and fern, and for
ever kissed by the little ripples which sped across the
crystal waters to cast themselves with a murmuring
sigh against these adamantine breasts. The eternal
lapping of the water as the ripples played among the
boulders, was the only sound that broke the strange
and almost weird stillness. Never a cry came from
the vastness of the forest, never a bird cast a mo-
mentary shadow upon the lake—all Nature, in fact,
seemed to be silent and inanimate.

A faint breeze was wafted over the lake, but not
enough to sway the boughs of the stern and rigid
pines and firs. The rippling waters sparkled brightly
in the sunshine, but I looked in vain for aquatic birds
upon their surface, neither was I able to meet with
much in the dense and solitary woods through which
I passed during my long tramp round the lake, but
in a thick spruce-wood I flushed a single "spruce partridge" or Canada grouse—a rather handsome grouse
which is said to be not very abundant anywhere in
this neighbourhood at the present time.

The strangest fact about these forests is that, in
spite of their lonely and retired nature, one is constantly imagining oneself in close proximity with frequented parts, or even habitations. Thus, whenever
I emerged into a more open part, I fancied I detected
signs of there being some track or clearing, although
the idea was always dispelled on closer observation ;
or again I frequently imagined I caught glimpses of
buildings through the trees, but if I endeavoured to
find my way to them, they vanished as completely as
the mocking mirage of the desert.

I was quite three hours on this lake, which at first
appeared to be by no means so extensive, although

continually opening up fresh arms and corners as I wandered on. I must have nearly made the circuit before I came upon a forest road to which I betook myself with some relief, and before long found myself on my homeward way.

CHAPTER V.

CTOBER 25.—To-day was Sunday, and relenting from his late fit of drowsiness the great sun-god shone forth in his splendour—flecking the ripples on the breeze-swept Bedford Basin with the golden light of daybreak and even bringing back a smile to the face of the dying summer.

Who could resist a stroll on such a morning as this, when even the bands of black-coated old crows indulged in hoarse laughs as they straggled over to the basin for a breakfast of fish or clams. Epicurean old rogues these crows are, enjoying the dainty " clam " as much as any unfeathered biped does.

As I approached by way of Three Mile House, a hawk sailing high overhead arrested my attention by the extent and curvature of its wings, and also by its loud and peculiar cry, which may be syllabled as *peetle-chu, peetle-chu*. It was a fish-hawk, or American osprey, a species which is very common in the United States, and also breeds sparingly on the coasts and lakes of Nova Scotia.

Up in the fir-woods I disturbed several American robins, an attractive bird and easily identified by its ruddy under parts ; their note was a low inward *chuck* or *chack*. I also observed a raven (*Corvus corax principalis*) near here flying over the road ; this form is almost identical with our European species, and is quite common in

Nova Scotia and also in other parts of Northern North America.

In passing I may mention that the great loon, or great northern diver, and also the red-throated diver breed not uncommonly on the Nova Scotian lakes, even within some few miles of Halifax. Both of these species are for the most part only summer visitors north of the United States, and I do not think they can be met with in Nova Scotia during the winter months.

The little pied-billed grebe, a bird belonging to a closely-allied group, is worthy of note as being the American representative of the familiar little grebe (*Podicipes fluviatilis*) of England. The American bird is just as widely distributed and abundant, while its habits are also very similar.

The great blue heron of North America bears no small resemblance to the common heron of Europe, to which it is in fact very closely related. It is a common and well-known species in the States, being as characteristic a bird there as our species is in England, although naturally not being met with in protected heronries as in the latter country. One morning, a few weeks previously, one of these herons—or " blue cranes," as they are popularly called—dropped down on a piece of marshy ground close to Halifax and not far from where I was living. But its temerity in venturing so near to the habitations of man cost it dearly, for it was soon observed and shot.

During a ramble in the woods around Melville Island on November 1, I observed several juncos, or black snowbirds, and some small parties of golden-crowned kinglets and chickadees. This last is a typical *Parus*, and has the crown, nape and throat black ; sides of the head and neck white ; mantle brownish-olive ; wing and tail-feathers edged with white ; and the under parts with a yellowish-buff tint, except the centre of the abdomen, which is white.

I also observed a single white-throated sparrow which crept away among the underwood uttering a slight chirp. This large and handsome sparrow has the head marked with a stripe of pure white in the centre of the crown, bordered on each side by a stripe of black, these again being succeeded by another line of pure white above the eye, with a yellow spot just in front of the latter; on the throat is a patch of silvery white from which the bird derives its name.

There were several crows about the woods, and I also came across a small party of American robins feeding on the ground in the woods. On being disturbed they flew up uttering a low *chuck* and settled on the bare limbs of the neighbouring trees, where they appeared to be reconnoitring; their call-note was a shrill whistle. The alarm note and the habit of flying up on to a bare branch reminded me very strongly of our redwing (*Turdus iliacus*), but systematists say that its nearest old world relative is the blackbird (*T. merulus*), and they have accordingly placed it in the sub-genus *Merula*, of which the latter species is the type.

The "migratory thrush," as it is also sometimes called, is a handsome species, and one of America's most typical birds. It leaves Nova Scotia about the end of October or the first week in November, moving southward to its winter quarters, and returning to the province again during the first week of April.

By the roadside I observed a fine specimen of the red admiral butterfly (*Pyrameis atalanta*); it settled upon stones with the wings expanded and was extremely swift in its flight. The American form of this insect is, I believe, regarded as racially distinct from the European, but the ordinary observer would certainly fail to detect any difference, either in appearance or habits. There is also a "white admiral" found somewhat commonly in Nova Scotia; it rather resembles the European variety, and I have seen it labelled as such in provincial collections, I believe, however, the species is designated *Limenitis arthemis*.

I visited the same woods again on November 12, and was fortunate in meeting with one or two Hudsonian chickadees in company with the golden-crowned kinglets. This chickadee is not nearly so common in the province as the black-capped species, from which it may be easily distinguished by its having the crown and nape warm brown instead of black, and the sides of the abdomen also warm brown, instead of pale buff, as in *P. atricapillus*. The name of this bird indicates its principal breeding range, and I do not know that it breeds in Nova Scotia, although it certainly does on Grand Manan Island, New Brunswick.

CHAPTER VI.

OVEMBER 14.—I visited, for the first time, the great woods rising from the western side of the Bedford Basin, and running back with but little interruption right into the heart of the province.

I left the Bedford road, which skirts the shore of the Basin, at a distance of some four miles from Halifax, and taking a narrow forest track, leaving the road at right angles, I ascended gradually until the great misty expanse of water lay far below me. Looking to the northward I could just discern the little town of Bedford, nestling in the valley where the river died away in the waters of the restless Basin; while out to the eastward I looked right over the harbour to the dim stretch of placid ocean beyond, glittering like burnished gold beneath the rays of the afternoon sun, and with many ships dotting the hazy horizon in the far distance. But I moved on again and soon passed the brow of the hill, and the fair vision faded behind the sombre stretch of forest.

The greater part of the timber here consists of spruce and hemlock, although other descriptions are interspersed in many places. In some spots the trees rise tall and slender to a considerable height from a swampy bottom covered with a dense growth of moss without a blade of grass visible, while in all directions lie limbs or trunks of trees in every stage of decay—some of them, although looking sound enough to the eye, being so decayed that in treading upon them one's foot frequently plunges into the heart. Very few birds were to be seen here; a crow at intervals sailed over, uttering hoarse croaks, and I also saw a red-shouldered hawk passing overhead with its somewhat irregular flight.

A little further on I came to a level bottom crowded with swamp-bushes, so tangled and dense that it would have been hopeless to try to force a passage through them if one left the cleared track. From its considerable extent and flatness, as well as from the fact that the trees ceased abruptly about midway down the shelving sides on either hand, I was convinced that a shallow winding lake had once existed here, and at no very distant time either, as was evinced by the spongy nature of the marsh or bog from which the mal-formed bushes sprang. Indeed through it still crept a sluggish stream, the waters of which were the colour of brandy from the masses of decaying leaves in its bed.

Beyond this again I came upon a high and dry sprucewood of astonishing density, the trees being crowded so together as to make it appear almost like the dusk of evening while I was in the wood. The sombre drooping branches of the spruce fir remind one of the plumes upon a hearse more than anything else, while their density almost effectually bars out the light of day, especially if the day should be a cloudy one. The ground beneath was covered with spruce-needles, with great boulders of granite protruding here and there and never a blade of grass visible. The stillness here was intense and almost oppressive, and was broken only now and again by the twittering of the small birds which occasionally came and settled in the trees, among them being a few black-capped and Hudsonian chickadees, and some numbers of golden-crowned kinglets. But I came across no larger birds and not even any animals, if I except a single red squirrel, which coughed and spluttered angrily at me as I approached its lurking-place in some low brushwood. Those who were not acquainted with this fearless little squirrel, would be

very properly disconcerted on hearing the peculiar
"whirring" or "rattle" emitted by it for the first time
while in these gloomy woods, for the sound commences
in close proximity to one and with startling suddenness,
giving rise, until the author of it is discovered, to unpleasant suggestions of "catamounts" and other unde-
sirable acquaintances which one is still likely to meet
with in this region. The "catamount," it should be
remarked, is the name by which the American wild-cat
is commonly known in the United States; it is not at
all rare in the forests of Nova Scotia, and individuals
are shot annually within ten miles of Halifax.

Leaving this somewhat gloomy region, I emerged
into a tract which consisted of smaller and more varied
growths of timber, and here I flushed a solitary ruffed
grouse, a bird belonging to the genus *Bonasa*,* and known
almost invariably throughout North America as the
"partridge."

I met an Indian towards Bedford, who told me that
he had recently killed two moose in the woods beyond
that place. The elk, or moose,† as it is called here, is
still, happily, not uncommon throughout the greater part
of Nova Scotia, but is never met with at the present
time within twelve or fifteen miles of Halifax. In the
more secluded forest fastnesses it still holds its own,
in company with the "cariboo," more familiar to most
people under its Old-World name of "reindeer."

This is particularly the case in the western portion of
the province, amid the great wilderness of mountain,
lake and forest, stretching from the South Mountains
bordering on the Bay of Fundy away to, and beyond,
Rosignol, the largest of the Nova Scotian lakes, secluded
and solitary, with its winding expanse of limpid, sun-
bathed waters, out of which rise clusters of miniature
islands sheer from its unruffled surface, upon which falls
the dark shadows of the spruce and firs, which have

---

* *B. umbellus togata* is the designation of the form inhabiting
Eastern Canada.
† *Cervus alces.*

stood guard for ages over the unbroken solitude of these fairy islets. Now and again, through the forest, one catches a glimpse of the barren but inexpressibly grand masses of the eternal granite hills, with the army of grim old firs and pines halting stubbornly half way up their rugged sides; baffled for centuries, but still with that hoary tempest-braving line presented to the foe.

The moose and the cariboo have from time immemorial shared this undisturbed wilderness with their sometime enemy, the Mic-Mac, who still makes this spot the scene of his hunting operations. Here, too, the bear, the beaver, the racoon and the porcupine are still found, while the American goshawk and the great white-headed eagle retreat into these fastnesses with the spoils of the chase, and here raise their broods secure from the gun of the farmer or the enterprise of the roaming egg-collector.

All along the coast stretching away to the south-westward of Halifax harbour, the great white-headed or "bald" eagle carries on his depredations. He is not quite such a "royal" bird as many writers have made him appear, unless indeed *kingliness* is not to be distinguishable from *despotism*. The white-headed eagle is at all times something of a despot. I think the favourite articles of his diet are fish and carrion, hence his partiality for the bays and estuaries of this wild and rocky coast. In connection with his liking for fish, Alexander Wilson's description of the bald eagle's spoliation of the osprey may occur to some. But he is also a great enemy to the poultry of the farmer, and has been often known to destroy young lambs, while he at all times commits great slaughter among the defenceless water-fowl, such as wild geese, ducks, the various species of gulls and other seabirds.

CHAPTER VII.

ECEMBER 6.—I received three fine female
birds of the American goshawk, which had
been trapped at Beaver Bank, a place
some distance north of Halifax, where
they had been making great havoc amongst
the "rabbits." In fact, when skinning
these individuals, I found that one had its
crop crammed with the flesh of the animal mentioned,
but the crops of the other two were nearly empty.
These birds appear to pass southward about this time
from their more northerly breeding limits, judging from
the frequency with which they are to be met with in the
province at this season.

This large and symmetrical hawk is one of the
commonest of the larger birds of prey in Nova Scotia
during the latter part of the year and again in the early
spring, and it also breeds in many parts of the province,
but I am not sure whether it is to be found there
throughout the entire winter or not.

Years ago, within the memory of Mr. Downs, this
fine species bred near Three Mile House, in the immediate vicinity of Halifax, but at the present day it
is not found breeding within, perhaps, twelve or fourteen miles of the city. In the great undisturbed tracts
of forest, however, particularly in the western portion of
the province, it breeds somewhat commonly. I have
examined several nests and eggs which were taken in

the region of the Blue Mountains and Gasperaux Lake, and in other places. The nest is usually built in a giant hemlock about the latter part of April or early in May ; it is a large structure, and composed of sticks, weeds, &c., lined with strips of bark and grass. The eggs are two or three in number, and in colour bluish-white, frequently unmarked, but sometimes faintly spotted with yellowish-brown. In size they measure about 2·3 inches by 1·8 inches.

Although not uncommon to the northward of the United States, this fine hawk seldom occurs within the limits of the latter, except as a winter visitor, and owing to its northerly range and inaccessible haunts, not a great deal is yet known concerning it.

The three birds previously mentioned were quite adult, and measured fully twenty-three inches in length by about three feet eight inches across the extended wings. The male bird, however, is considerably smaller. This species differs chiefly from the European goshawk (*A. palumbarius*) in having the upper part of the head deep black, with a partly-concealed patch of white on the nape, and the feathers of the entire lower parts of the body marked transversely with many fine zig-zag dusky lines, each feather also having a fine black central streak. The European goshawk, on the other hand, has the under parts thickly and distinctly barred with ash-brown. The immature bird in the American species, as in the European, has the plumage of the lower parts boldly marked in a *longitudinal* direction.

I also found the strikingly-marked harlequin duck to be fairly common in the province about the same date, and I saw many of these birds brought in during the early part of December.

Some black guillemots were also brought in. By this season of the year the deep black summer plumage of the adult had changed somewhat, the crown being white marked with black, the mantle barred with black and white, and the lower back and under parts of the body almost white. This species breeds commonly on the coasts of Nova Scotia and New Brunswick (notably

on Grand Manan and other islands in the Bay of Fundy)
and from thence northward to Greenland. It retires
southward in winter, as far as Massachusetts, but
appears to remain on some parts of the Nova Scotian
coasts throughout the winter.

Up to December 15 the weather was generally
fine and very mild indeed, except for sharp frosts at
night during the previous week, but on the night of
this day snow fell, and lay about two inches deep in
the morning, although it disappeared during the day
in consequence of the temperature changing and rain
falling rather heavily.

On the 21st I noticed that the "rabbits" were nearly
white ; it is curious how quickly after the first fall of
snow they seem to assume their winter dress.

Owls appear to be common in Nova Scotia about
this time of the year. On this day I noticed about
the town, and in the taxidermists' shops, both dead
and living specimens of the great horned, snowy,
and barred owls, all which had been taken in the
province. I also saw a fine stuffed specimen of the
great grey, or cinereous wood owl, a huge species,
exceeding in size both the great horned and snowy
owls, and even its near relative, the great Lapp owl
of Northern Europe. In common with almost all
Arctic birds, it is somewhat soberly attired, the plumage being mottled and barred with white, brown and
grey. This individual had no doubt been taken within
the limits of Nova Scotia, as the species even occurs
within the boundary of the United States during the
winter. The summer home of this great owl is in
Arctic America. Dr. Richardson met with it commonly
on the borders of Great Bear Lake, and records that it
keeps constantly in the woods, hunting hares and other
smaller quadrupeds.

On the 27th I revisited the woods in the neighbourhood of Melville Island. There were not many birds
about, but I observed some few brown creepers, which
little bird remains in Nova Scotia throughout the
year, as its Old World brother does in England,

although I have no doubt that part of the birds to be met with during the winter have migrated from more northern breeding haunts, as is also the case in the British Islands, where immigrants from the continent swell the number of our native birds in the winter.

The handsome and well-known blue jay seems to be less common than formerly in the neighbourhood of Halifax. Once or twice only have I heard the somewhat harsh squall of this bird in the woods around Halifax, although I have heard of as many as six or eight having been seen at one time within two or three miles of the city.

The blue jay is certainly one of the most beautiful and entertaining of the commoner American birds, although at the same time one of the most wary, possessing in fact many of the characteristics of our European species. The present bird is rather smaller than the latter, and has the head handsomely crested with loose silky feathers ; the upper parts of the body of a fine blue colour, deepest on the wing and tail-feathers, which are barred with black and tipped with white ; the neck is encircled with a black collar, and the throat is whitish, tinged with blue.

According to some writers the blue jay is an expert mimic, but for my own part I prefer to believe that the mimicry exists more in the imagination of the said writers than in actual reality. Indeed, I would almost venture to say the same of all the other birds which are so commonly said to mimic the songs and notes of other birds and animals. I will not, however, express an opinion upon the mocking-bird's talents, as I have not yet heard the song of that bird in a state of nature. But I can distinctly assert that not a single one of our British birds is guilty—in its natural state—of *consciously* imitating any part of the note or song of another species. Perhaps the sedge-warbler, among our native birds, has most frequently had these powers of mimicry attributed to it, but, although as familiar with the song of this bird as with the chirpings of the sparrow in our streets, I can positively assert that I never yet heard it

3

utter a note that had more than a slight resemblance
to that of another songster, and I am therefore persistently incredulous with those who detect in its song
the borrowed notes of other birds, inasmuch as I
maintain that such resemblance is more accidental
than real.

There is another species of jay found in Nova Scotia,
but much less frequently than the last-mentioned. It
is the Canada jay—also known as the "moose-bird,"
"whisky-John," &c. This jay is a resident species, but
is apparently not at all common in the greater part of
the province, and certainly not so in the neighbourhood
of Halifax. The bird is about the size of the blue jay,
but of much more sober plumage, having the upper
parts of a leaden-grey tint—with the exception of the
crown, which is blackish—and the under parts dirty
white, the tail being long, cuneate, and tipped with dull
white. The whole plumage is long, soft, and loose, and
no doubt forms an ample protection against the severity
of the Canadian winter.

CHAPTER VIII

ANUARY 2.—Overhead the dim blue of a cloudless sky, with pale stars fading from the dawn, and the great fiery sun peeping over the hills, while every tree and post upon the wayside loomed weirdly through the hurrying mist. Beneath my foot the snow crunched sharply, every crystal sparkling with many colours in the first struggling rays of the rising sun.

Soon the mist rolled in great banks across the fields and filled alone the hollows, and out of these came an early crow or two, flapping leisurely along with an occasional querulous cry, and head turning from side to side continually on the look-out for plunder.

As I neared the North-West Arm, a solitary sleigh glided past me, the horse's trot falling softly upon the snow to the pleasing accompaniment of the tinkling bells. Crossing the head of the Arm, I turned to the south-east, ascending the road to the hanging forest of hemlocks above, standing out darkly through a faint enveloping mist, tinted with a soft light by the morning sun.

Presently I left the road and plunged into the solitude of the fragrant pines and hemlocks, stepping over the virgin snow that lay, crisp and glistening, fully five inches in depth upon the ground. There were a few wary old crows about these partially cleared outskirts of the forest, calling to one another from the tree-tops with a loud imperious note, much like that of their Old World cousins. They appeared to be at least as wary as the latter, and it is difficult to shoot one of them, while residents in this part will seriously assure you that it brings " bad luck " to kill one of these birds.

Among the pines here, I met with a pair of red-breasted nuthatches, a bird which has much of the habits and appearance of our familiar species. I noticed that these birds crept about upon the lesser limbs of the pines, usually the higher ones, and I did not hear them utter any kind of note. Like other small birds at this season, they traversed the woods from one tree to another, with but little stay upon each, and I soon lost sight of them. Now and again I observed a few crossbills or grosbeaks (I could not be certain which) at work upon the cones in the summits of the taller trees. I also heard the sibilous note of the brown creeper, and detected a pair of these little birds creeping up the trunk of a small fir.

In spite of the cold I noticed some few of the small red squirrels about soon after entering the forest. Unless alarmed they were quite fearless, and would call one another from the branches close to me with a curious little cough or bark, also squeaking and chattering :—

> " Sprang the squirrel, Adjidaumo,
> In and out among the branches,
> Coughed and chattered from the oak tree."

On the whole, however, very little life was to be observed about the woods during this long ramble, sometimes through great tracts of forest with nothing but the crunching of the snow under foot to break the strange stillness, or across densely-wooded bottoms strewn with great granite boulders now covered with four or five inches of snow, but beneath this green with the luxuriant growth of moss, while after a time I gained the higher land, open and scrub-covered for the most part, but with many little hollows filled with a dense growth of young hemlocks and firs dim with the faint mist of a January day, clothed with long tufts of tree moss and decked with white and green lichens, recalling forcibly to the mind Longfellow's beautiful lines :—

> " The murmuring pines and the hemlocks,
> Bearded with moss and in garments green, indistinct in the twilight,
> Stand like Druids of old, with voices sad and prophetic,
> Stand like harpers hoar, with beards that rest on their bosoms."

To me there is nothing wearisome in these woods. I am, indeed, far more lonely and depressed among the habitations of man and the bustle of his selfish avocations than in the most secluded forest, for in these great wilds the spirit of Thoreau and of many another brother in Nature, rises up to bear one company, filling the mind with an absorption and obliviousness to care that I have tried in vain to attain to within the sordid limits of my chamber. And then the birds and beasts that one follows on unweariedly into their most sacred retreats, as they move restlessly through the forest, intent upon their own pursuits and careless of your stealthy approach and engrossed, untiring observation ; squirrel and crow, nuthatch and tree creeper, all alike oblivious of your silent intrusion and even your very existence, while they gambol, and court, and feed, each in its unconcerned way, until the tyrant—biped or quadruped —puts an end to their harmony, and often also their existence.

How puerile and foolish it seems to presuppose that aught of real *beauty* or *peace* exists beneath the pleasant sunshine, the sombre undulating forest, and the virgin-white robe that Nature wears to-day! Always amid this semblance of peace and gentleness, and these timid birds and animals, or the insects of the past summer, lurk the ever-recurring agony and the violent death. How vain, indeed, it is to think of *peace*, in a system whose harmony and balance rest solely upon Utility— the Survival of the Fittest!

CHAPTER IX.

N January 17 I visited again the woods around Melville Island, but saw very few birds beyond several " partridges " and chickadees—or black-capped titmice—and also the ever-present crow.

On the 24th I observed snow-buntings for the first time near Halifax. No doubt the recent great snowfall and severe cold had driven them southward, as they usually make their appearance in these latitudes after the first severe snowstorm. Nothing can exceed in interest the sight of a straggling party of " snowflakes " drifting across the snow-covered fields, or sometimes flying close over the road with their wavering flight and feeble twittering notes.

I observed some small parties of golden - crowned kinglets in the woods again on January 30, and since I had failed to detect any of these birds during the previous six or seven weeks, I had good reason to conclude that they had moved further to the southward during that period, to escape the severe weather.

Up to February 16 small flocks of snow-buntings were still about the fields in the vicinity of the town, picking up a precarious livelihood from the dung-heaps and refuse of all kinds, also paying great attention to the seeds contained in the crowns of the few withered plants which projected above the uniform stretch of snow in the fields and along the ditches.

The sight of these little fugitives from the rigours of a Polar winter provides mental pabulum for many strange reflections, and so one finds oneself thinking *why* these small frail creatures—to whom most would deny the possession of reason—why and how they should travel southward so unerringly to these less ice-bound climes ; and why—still less explicable—they should return as unerringly to those same Polar regions when the fitful Arctic summer sets in—nay, even before, for they depart from their temporary refuge long before the advance of genial spring.

Years ago such questions as these were left unargued, and were, indeed, unanswerable, but now we have learnt somewhat of Nature's secrets, and can reason back to the time when a warmer climate reigned over the far north, and the birds were stationary, as are those of more temperate latitudes at the present day. Then, as the great winter, or ice age, set in, one can picture the whole animal life of this great area being forced gradually southward with the extension of the cold region, while in the brief spaces of the Arctic summer this southward movement would naturally cease for a time ; the frost-bound trees would perhaps struggle forth into leaf again ; the dormant insects would again appear ; while the birds, by reason of their powers of flight, would even venture to return somewhat towards their former breeding haunts. This may be only hypothesis, but it seems to clear away much of the shadow of the inexplicable surrounding the question of migration.

A few northern shrikes or " butcher-birds " came under my notice during the month of February, and one day a bird of this species was actually to be seen in one of the principal streets of Halifax, to the no small terror of the over-confident alien sparrows, one at least of which it pursued and captured. This bird is closely allied to the grey shrikes of Europe, and possesses similar habits, including that of impaling its prey upon thorns.

There were also in the vicinity of the town at this season a somewhat unusual number of " saw-whet " or

Acadian owls. This owl, the smallest in Eastern North
America, is quite a common species in the province of
Nova Scotia, particularly at the present season.

By the 28th the snow-buntings had all disappeared,
the weather being milder, and on this day I observed the
black snowbird again in the woods by the North-West
Arm.   I also noticed a flock of shore larks—" horned "
larks in New-World phraseology—circling over the
fields; they flew in a very compact troop, and appeared
to number quite one hundred.   Crows, I noticed, were
already in pairs, although not nesting here until the
latter part of April.

March 13 brought very little change in the weather,
the snow still lying ten or twelve inches deep in the
woods, while the only birds I took notice of during a
brief ramble were a few chickadees, one of which I disturbed as it was busily engaged at a bunch of pine-cones
which had lodged in a bush.   I observed, however, that
the squirrels had already commenced to prepare their
spring nests in the fir-trees.

On the 26th, also, the weather was unaltered—
although not particularly cold—the snow still lying un-
diminished in the woods.

CHAPTER X.

HE first of April opened the month well, but the succeeding day was even finer and warmer, and in consequence the snow had already quite vanished in the fields, and was rapidly diminishing in the more open parts of the woods. Indeed, in sunny little glades, where the snow had all melted, I several times startled hybernated individuals of the Camberwell beauty butterfly (*Vanessa antiopa*), an insect which is very commonly distributed in Nova Scotia. They almost invariably settled upon the ground, with wings outspread to catch the full warmth of the sun's rays, starting up suddenly upon one's approach and flying off with extreme swiftness.

I disturbed several chickadees from an old decayed stump of a tree, about ten or twelve feet in height, in which I found they had commenced several holes, probably with the intention of forming nesting cavities. Some numbers of crows are always to be found in the morning in the woods upon the North-West Arm, along the shore of which the ground beneath most of the trees is strewn with clam-shells, proving the extent to which this favourite bivalve enters into the "bill of fare" of these omnivorous birds.

In several parts of the forest I found a few fox sparrows —a large and handsome sparrow, the largest in North America, and noticeable from its ruddy tail-coverts and tail, and the bold blotches or markings upon its breast and flanks. I met with the first party of four or five of these interesting birds among the rocky woods near Melville Island, and followed them back into the woods for some distance, while during the remainder of my ramble I several times came upon a few of these birds— always in the most secluded and dense woods of fir and hemlock, and generally in a spot where rocks or boulders appeared. Although of such shy and retiring habits, they seemed rather unsuspicious, and allowed a some- what near approach unless alarmed. I often saw one upon the moss-covered boulders, apparently seeking for insects or their pupæ, while I sometimes approached quite close to them when perched in the trees, and I noticed that they always settled low down on a dead branch or bough, against the trunk, where they sat so motionless that they might be easily passed un- noticed. The only note I heard these birds utter, and one which seemed common to both sexes, was a low and somewhat plaintive whistle.

In the fields nearer the town I several times noticed the familiar chipping sparrow. This sober little bird usually sang from the topmost spray of a bush, the song being a weak, but not unpleasing, little trill.

The following day (April 3) was cloudy in the morn- ing, but cleared up before noon, and was very fine and warm the rest of the day. I walked over to the Bedford Basin in the morning, and from thence round to the North-West Arm, the walk being thoroughly enjoyable, although not much of interest was to be observed on the way. In the woods by Melville Island I came across some newly arrived " robins " or migra- tory thrushes, and also observed a pair of the prettily- marked pine-creeping warblers upon the trunk of a huge pine, while in the trees and bushes close by were a few " yellow-rumped " or myrtle warblers, this species being about the earliest to arrive of the many warblers which visit the province. They mostly frequented the bushes or small trees, and often uttered a simple, pleasing warble.

I heard a chickadee here uttering a note or song

almost identical with that of our great titmouse; it consisted of the three notes only, however, there being no quick repetition of them as in our bird.

Just before dusk I noticed fourteen geese—probably Canada geese—flying over the town from west to east at a considerable height, and apparently following the coast line. My attention was drawn to them by their occasionally-uttered cry, a deep *honk, honk*, and they were then flying in straight single file of twelve, with two on one side abreast, but soon afterwards they opened out at this side into an irregular V-shape.

It was on this night, also, that I witnessed a beautiful aurora to the northward. It lasted several hours, during which period the faint shifting rays of light illumined the whole sky to the north and north-east, producing at one time a most striking and beautiful effect.

The morning of the 15th opened very dull and cloudy, and there was not a breath of wind to be felt, while there seemed to be a promise of snow later in the day. I left home soon after daybreak, and after two or three miles' walk struck the woods near Three Mile House, on the Bedford Basin, from whence I passed through continuous dense woods in a north-westerly direction all the morning.

Soon after entering the woods I came upon a party of golden-crowned kinglets in a dense undergrowth of young firs, and except for these there was hardly a bird to be found in this part of the forest. Some of the firs here were of a prodigious size, and near the summit of one I noticed what appeared to be an unfinished nest of a hawk or crow.

A little farther on I came to a small sluggish stream flowing through a swampy hollow. The brandy-coloured water was fringed upon either side with curiously-contorted swamp bushes, while here and there a fallen trunk bridged the stream; but I had learnt by experience not to trust these seeming bridges, for although looking sound enough to the eye, they are usually mere shells from which the heart has long since

rotted away. I noticed, as an odd circumstance, that
for some distance this little stream formed an abrupt
boundary between two totally different descriptions of
woodland, for while upon one side the woods consisted
of (at this season) bare and leafless birch and similar
trees, on the other side of the stream rose dense and
funeral-looking spruce woods. I heard a woodpecker
tapping here, and followed it a short distance, but could
not obtain a view of it. I also noticed a white-breasted
nuthatch among the firs, as well as several chickadees
and brown creepers.

Leaving the stream I then struck through the forest
until I came upon a long ravine in the dense woods,
with sloping sides and an almost level bottom, which
was sparingly timbered, and with a sluggish stream
winding along it, and on which the snow still lay thickly,
as it also did in many spots in the surrounding woods.
I saw here a large nest of the American crow, fully
sixty feet up in the fork of a large and almost limbless
maple, but I did not attempt to ascend to it. Crossing
the ravine I pushed on again through very thick woods,
varied occasionally by higher and more open rocky
ground covered mostly with scrubby brushwood, but
saw nothing more, and so retraced my steps and struck
a track leading back to the Bedford Basin. Just here
the trees were chiefly hemlocks, and veritable giants of
the forest they were, many of them being fully four feet
in diameter near the ground and towering to an immense
height.

While passing down this track the snow commenced
to fall gently, adding to the lonely and desolate nature
of the forest. I noticed many old burrows of the
woodpecker in the summits of the bare and whitening
trunks that met the eye on every side, but saw very
few birds of any kind until I came to a low swampy
fir wood, in the midst of which was a shallow pond
where the frogs were croaking dismally. Here a large
barred owl flew close past me with its peculiarly light
and noiseless flight, and settled upon a dead tree a
short distance behind. The habit of flying in the

daytime in this species is well known; it was shortly
after noon when I saw the present individual, and snow
was falling rather fast at the time. I judged it to be
a male, the female being considerably larger in size,
and in fact almost rivalling the great horned, or
American eagle-owl.

Continuing on my way I disturbed a pine grosbeak
from the ground at the foot of a fir-tree; this is a very
handsome scarlet-tinted bird and is by no means
unknown in the province at certain seasons, although
breeding farther north. Pushing on again, the great
foggy expanse of Bedford Basin soon came into view
below me, and before long I reached the road which
winds round by the water's edge, and started homeward,
well satisfied with my day's ramble.

On the following day I came across a solitary pair of
fox sparrows in a rocky and elevated spot by Melville
Island. These were evidently very late stragglers, as
the majority must have departed to the northward ten
days before, this sparrow breeding farther north, in
Newfoundland and Labrador.

At about 11 a.m. on the 25th I was in the woods
near Three Mile House when my attention was attracted
by a great outcry among a party of perhaps a hundred
crows wheeling over the tree-tops at a short distance.
I soon perceived that they were vigorously mobbing
a large barred owl which was sailing leisurely along in
the direction of the Bedford Basin, the crows closely
surrounding the stranger, and darting down so close
as almost to touch it, uttering all the while loud and
incessant outcries. The owl seemed but little concerned by their attacks, however, only occasionally
uttering a low harsh scream or growl, while it sailed on
straight ahead, soon leaving the majority of its per-
secutors behind and being only pursued by four or five
of the crows, which followed it right over to the great
woods across the Basin. No doubt the excited resent-
ment of the stranger's visit by the crows was largely
due to the fact that the latter were actively engaged in
nesting operations.

It was late in the afternoon as I returned home-ward through the woods near the lake at Three Mile House, and quite a number of robins were singing in the tree-tops at the outskirts of the woods. Their song is loud and possessed of little variation, but still attractive; it is certainly inferior in mellowness and compass of voice to that of the vocalist's Old World cousin, the blackbird. The song may be readily syllabled as *gie-it-up*, *gie-it-up*, *gie-it-up*, *pilly*, *pilly*, but it is strange what an amount of rivalry and assertion it conveys, for the birds will sing one against the other with a surprising vehemence and vigour for an hour at a time.

In a shallow grass-grown pond which I passed before leaving the woods the frogs were holding a merry concert. Heard in the twilight, in the stillness of the forest, there is something plaintive in their clear and shrill *peet*, *peet*, uttered at first by one only and being every time answered by another and another, until all join in one swelling chorus—

> " And anon a thousand whistles
> Answered over all the fen-land."

CHAPTER XI.

AR different to all our old-time conceptions is the dawn of a New World May·day in the solitudes of the primeval forest! No groups of villagers, no merry dances, no gaily-decked teams of horses—nothing but the grey silence of day-break, and the all-extending forest.

As I stand, the woods close in around with their array of shadowy forms looming through the uncertain light of dawn. A space further on a low boulder forms a ready couch. Here the ghostly army of the forest fades away, for below is the sea, now lying placid and dumb, with a faint slow heave of its fair bosom, and a mute, passionless appeal which draws one's thoughts out to it and steals them away seaward —over to that Old World from which the face was so resolutely turned.

These are the moments of reverie, undisturbed by any sound save the ripple of the tiny waterfall near at hand. Here it is always water—little streamlets splashing from every hillside and chasing one another down among the hollows and shallows and the littered granite ; down, down, and away headlong to the sea—

"Run home, little streams,
With your lapfuls of stars and dreams "—

singing, as they run, little intermittent snatches of strange music ; now like the faint, far tinkling of silver bells, and again like the sedge-bird babbling by the brook in |far-away England. But it is not *peace* they sing of, not simple contentedness, for into these reveries comes a great, passionless unrest, born of thoughts of a breadth and scope far too subtle for the mind to grasp, as sweeping and as unabiding as the ocean that frets away for ever upon the granite below, and as un-trammelled by the narrow injustices and heart-searings of the passionate world of man.

But a faint rare flush upsprings in the pale east, deepening to a glow, ascending to the zenith, and over-spreading until but a lone pale star twinkles low in the west—a jewel glistening upon the train of flitting Night, until up from across the sea peeps a great fiery disc of dazzling gold, flashing forth the triumph of spring over earth and sea and sky, and waking to busy life the countless denizens of field and forest.

Already the feathered choir have been heralding the coming of the great Life-giver, and the woods are astir with song—the petulant whistling of the " robin," the chant of the water-thrush, and the many trillings of the warblers, while along the shore the sable crow swells the symphony with discordant music. But anon comes another and a stranger song from the shady recesses of the underwood—notes sweet-toned and changing, like the babbling of a brook over stones, and with as sad and stately an undercurrent. It is the song of the hermit thrush—bird of the dark and gloomy forest, the secluded swampy hollow, girt round with dense underwood and crowded with tall breathless firs, staring up for ever from out of an endless twilight—

> " From deep secluded recesses,
> From the fragrant cedars and the ghostly pines so still,
> Came the carol of the bird."

     *   *   *   *   *

The first of May seems to commence the spring in Nova Scotia, judging by those sure guides, the birds: and indeed the day was so delightfully fine that I felt justified in honouring it as the first day of spring. The trees, however, had scarcely commenced budding as yet, although the snow had all vanished in the woods.

Down by Melville Island I came across five or six nearly completed nests of the American robin—none of them, however, containing any eggs. One was placed at the extremity of a branch high up in a pine, about thirty feet from the ground, this being the greatest altitude at which I ever saw the nest of this species. Another was straddled on the thick horizontal bough of a hemlock, at a height of not more than eight feet. A third was cradled in the drooping branch of a hemlock, and so low down that I could touch it with my hand; while one was even placed in the fork of a bare silver birch. Both in situation and appearance the nests closely resembled a neat example of the European mistle thrush's nest.

The robin, or migratory thrush, is the first to commence breeding among the smaller species of birds found in the province, and as the period when it commences nest-building corresponds exactly with that of the Old World blackbird and mistle thrush, it may be readily imagined how much later the spring commences in Nova Scotia, in spite of the fact that it is several degrees farther to the southward than England.

In the same woods I disturbed a downy woodpecker from a decayed fir some eight or ten inches in diameter. It allowed a very near approach before revealing itself, when it darted on to another stem a couple of yards away and clung motionless to the bark, peering down at me for a moment or two, and then flew silently away. I found that two excavations had been commenced in the tree the bird was disturbed from, both of them being about twelve or fifteen feet from the ground, and being scarcely larger than those made by our lesser spotted woodpecker, but they did not penetrate more than a couple of inches.

The downy woodpecker—sometimes inappropriately called the lesser "sap-sucker"—is a very small species, in fact, the smallest in North America, being not more than six inches in total length. It is noticeable from its black upper plumage and white lower parts, contrasting with its scarlet hind-head and crest. This

4

little woodpecker has a curious habit of drumming continuously upon a resonant part of a decayed limb, probably with the object of calling the attention of its mate, for when it ceases for a while an answering tapping may be plainly heard proceeding from another direction.

On the morning of May 7 I observed the first swallow in the neighbourhood of Halifax; it was of the species known as the white-bellied or tree-swallow, and distinguished by the entirely white lower parts.

During a ramble in the woods by the Bedford Basin I was much amused, while taking a brief rest upon a fallen tree, by the actions of one of the little red squirrels, which scampered along the ground around me, and sometimes came close up under the cover of the brushwood, uttering its shrill and fairly startling rattle, and coughing, spluttering and whining in a most ludicrous manner, as if in a great rage at my intrusion upon its retreat.

The troublesome "black fly" (*Simulium molestum*) a dipterous insect allied to the gnats, makes its appearance in unpleasant abundance at this season. It appears, however, to be entirely confined to the more secluded and damper parts of the woods.

In a rather open part of the forest I came upon a nest of the "flying squirrel" (*Pteromys volucella*) in a young pine at a height of about twelve feet. Both the squirrels were at home, for on tapping the nest they left it and made their way silently up to the branches above, where they remained clinging mute and motionless to the limb in such a manner as to almost escape notice, owing to their resemblance to a patch of fungi, or other foreign matter. Their habit, indeed, is to remain at rest during the day, and to come forth only after sunset, when they may sometimes be seen gliding swiftly through the air from one tree to another. They are very pretty little animals, about the size of the common red squirrel, and of a light brownish, or rather drab colour above and silvery white below, the fur being soft and silky and very close. The animal is furnished with an extensible fold of skin on either side, from limb to limb, which is stretched tight when the limbs are extended in leaping through the air ; the extension is aided by one of the toes of the fore-paw being much lengthened and inclined backward ; the tail also is broad and concave below.

The trailing arbutus, the exquisite little "mayflower" —emblem of Acadie—is in full flower about this time of the year, and sure enough during the day I found it growing in great profusion along the side of a forest track. The pretty little white flowers, half hidden among the leaves, remind one somewhat of our white violet, and it would be hard to decide which possesses the sweetest scent. All the esteem in which the violet is held by Old-world Nature-lovers is lavished upon the little "mayflower" here—and in much the same way, for each spring it is sought for diligently, torn from its enfolding leaves and carried heartlessly away to droop and die in a glass or vase in some forgotten corner.

But as with flowers so it is with birds. These beautiful denizens of field and forest have no enemy more to fear than *civilized* man, who is ever on the watch to slaughter —ever inventing fiendish appliances to capture and cage them. Better, a hundred times better, were these celestial pipers stretched stiff in death by some more natural foe than doomed to pass their innocent lives cooped up in small wired prisons far away from the haunts of their delight, to sing their plaintive strains day by day into the deaf ears of a heedless and uncaring crowd. Strains of sorrow and upbraiding are they as summer fades and the little prisoner longs for the clear sweet air and his kindred's society among the quiet upland stubbles. Dreary songless days of winter, amid slush and fog, and far from his native woods and meadows, now covered deep with crisp sparkling snow, every little coppice glistening with the fairy forms of a thousand enchanted palaces rising turret upon turret and spire upon spire, desolate and silent in the stillness of morning ; until perchance the sable blackbird breaks the spell as he darts out noisily, scattering a million sparkling gems in his flight, and chasing away the fairy

vision with his hearty laughter. So spring comes, and
from his narrow cage the poor imprisoned songster
pours forth incessantly his tale of love to far-away ears
that hear not, straining his soft breast against the cruel
bars that alone separate him from life and liberty.
And so the spring passes, with its gladness of warmth
and sunshine, its busy birds, and insects, and flowers,
while the little captive ceases his song, and, perchance,
languishes gradually away, and before spring returns
Death has ransomed him from his troubles—his all-
powerful touch has burst the narrow prison and carried
away the frail remains to winnow in his mighty granary
and plant again in the garden of Life, who will water
them with living tears until the *eternal spring* shall raise
them once more into some fair blossom that will go
forth afresh to scatter seeds of truth and beauty upon
the earth.

CHAPTER XII.

HE next day found me again in the woods by Melville Island, and on inspecting the robins' nests I discovered that the birds had commenced to lay, there being three eggs in two of the nests, two in another, and one having four. This last nest was the neatest of them all, its situation being the drooping branch of a small hemlock. It was not so bulky as the nest of our mistle thrush, but similar in construction, the exterior being constructed of small twigs and tassels of green tree-moss (*Usnea*), while the lining was a neat cup of fine dry grasses ; I noticed also that the nest was secured to the branches with a small quantity of moist loam, while there was the usual intermediate wall of the same in the nest itself. The eggs of the " robin " are unspotted, of a uniform and rather deep greenish-blue, measuring about 1·10 by ·85 inches.

By May 13 the ruby-throated humming-bird had already accomplished its wonderful journey to the province, as I was shown one which had been sent from Kentville on this date. I was informed that the previous spring one of these beautiful little creatures visited the flower-box in one of the windows of the house I was stopping at and was observed hovering there for some minutes.

Fine weather on the 14th again took me to my old haunts on the Melville Island inlet, and in the woods here I noticed some more nests of the "robin," but very few other birds seemed to be interested in nidification as yet. In a little open spot I came to a decayed maple stump, about twelve feet high, near the top of which a pair of "flickers," or golden-winged woodpeckers, had been very busy excavating a suitable burrow for nesting purposes. The cavity within was large and roomy, being about ten or twelve inches deep, although apparently not yet completed, while the whole ground for yards around was littered with the chips and dust in evidence of the arduous undertaking.

This habit of building has earned for the bird the rather curious appellation of "high-hole" in some parts of the United States. It is under this name that Whitman, a true naturalist (as every real poet needs must be), mentions it :—

> "Put in April and May, the hylas croaking in the ponds,
> Bees, butterflies, the sparrow with its simple notes,
> Blue-bird and darting swallow, nor forget the high-hole flashing his golden wings."

Near here I visited a nest of the flying-squirrel, which I discovered a few days before in a hollow in a small stump and only four or five feet from the ground ; there were then three young squirrels, blind and naked, and the female in the nest, but I found on re-visiting it that the squirrel had removed her young from the nest, no doubt to a safer hiding-place.

Passing from here by a slight track through some rather open rocky ground, I startled a nighthawk from the ground close to the path. Like its English cousin, the nightjar, the bird harmonises well with its surroundings, and also crouches so still that it is never seen until one's close approach rouses it to flight. In general habits, too, in flight and in nidification this bird of the twilight is almost a counterpart of its European relative.

The next morning awoke me with the sounds of spring—the little alien sparrows outside my window quarrelling over the possession of some disputed nesting-place—and I arose to look upon a day of singular glory, even for this land of clear skies and lovely weather.

The great charioteer of light had already mounted high in the cloudless heavens when I set out, and everywhere I found the splendour and the fulness that go to stamp such a day as this in indelible characters upon the memory. Going down to Three Mile House through fields just awaking to the influence of spring, and laughing with the multitude of song, I saw many birds. Some swallows flew above me, one the barn swallow, much like that of England but differing in the arrangement of its colours. In the fields were many sparrows, for Nature has endowed the New World with a multitude of these, while man has added yet one species more—the " house sparrow "—and that one now more maligned than all the others put together.

Down in the valley the fields terminated in a wet swamp, beyond which I came into some woodlands. Here it was that a small speck I had been watching resolved itself into a little hawk hovering in the air, sometimes dropping down a little, then rising up with a circling flight and again hanging quite stationary except for the slight vibration of its wings, exactly after the manner of our familiar kestrel or "windhover"—to which, indeed, the American sparrow-hawk, as this little falcon is termed, is closely akin. It is in habits and appearance almost an exact counterpart of our kestrel, although much smaller — in fact, scarcely exceeding the size of a thrush, the female (which is the larger) being not more than eleven inches in length. It preys largely upon field-mice, lizards, and various insects, often also upon small birds. The immortal Wilson mentions having taken from the crop of one of these little hawks a considerable part of the carcase (including the unbroken feet and claws) of an American robin, although the latter is scarcely smaller than the dashing little freebooter itself! The eggs are laid within hollows and holes high up in the trunks of trees or in the crevices of rocks and cliffs, but seldom is any nest constructed. Eggs that I have seen have been quite diminutive, with a creamy or pinkish-white ground marked boldly with reddish-brown and cinnamon colour, chiefly around the larger end, but they vary much in appearance ; the average size is given as 1·36 by 1·12 inches.

Passing through a little hollow near here my attention was attracted by a sudden outcry among the small birds in the underwood, and next moment out darted another of these little hawks in front of me, its bright reddish lower-back being conspicuous as it flew away, and proving it to be a male.

The warmth and splendour of the day seemed to have aroused to life the whole insect creation, for in the cleared spaces of the woodland grasshoppers stridulated upon every hand, while flies—noxious and innoxious—were everywhere. Across these little openings, too, came every now and again small troops of a pretty little *Polyommatus*, or " blue " butterfly, dancing over the ground like a " will-o'-the-wisp," and vanishing down the openings of the woodland. A small *Thecla*, or " hairstreak," was also not uncommon here, flying close to the ground, and from its small size and dark colour being easily overlooked, especially among the small scrub, like the heather of an English moor, with which the ground was covered. Once, too, a still-surviving " Camberwell beauty " scurried past me and hastened away over woodland and swamp, while among the underwood I now and again noticed small moths of various kinds upon the wing. Emerging upon the lake by Three Mile House I met with one of our common white butterflies (*Pieris rapæ*) and also saw another shortly afterwards. This butterfly was introduced into the United States about the year 1866, but I understand that it was unknown in Nova Scotia for some years after that date, nor does it appear to be very widely distributed there at the present day.

On the 17th I visited again the woods running back from the Bedford Basin, but, strange to say, although the day was fine enough there was exceedingly little of

interest to be met with. Indeed, spring appeared to be
a week or two behind in these great desolate forests as
compared with the fields and woodlands adjoining the
city. On the outskirts of the forest I noticed a few
white-throated sparrows, and also one or two golden-
crowned kinglets, while on reaching a little swift-
running brook I met with a single black-and-white
warbler, which attracted my attention by its pleasing little
song, which might be syllabled as *chiv-vee*, *chiv-vee*, repeated in quick succession. The bird itself frequented
the topmost branches of the trees, occasionally creeping
upon them or around the trunk in a similar manner to
the tree-creeper. As its name implies, this little warbler has the plumage prettily variegated with black and
white. This was indeed the only bird of interest met
with while in these woods, but in a dense little tract of
young firs I found a nest placed against one of the
stems at a height of about eight feet, which most
probably belonged to the olive-backed thrush, a common enough bird in most parts of Nova Scotia. This
nest was shallow and loosely constructed of moss,
leaves, twigs, bark strips, &c., but it contained no eggs.

Fine weather again on the following day took me
along the St. Margaret's Bay road, turning off to the
right from the head of the North-West Arm, and passing
by Chocolate Lake. It was as I neared the latter that
I first observed a pair of belted kingfishers, which were
circling round and round high above the lake uttering
a loud harsh rattling note, and altogether presenting
very little resemblance to our English species, which
they also considerably exceed in size. I never observed
a kingfisher in Nova Scotia during the winter, so that
it is evidently a summer visitor to the province. The
same may be said in regard to the golden-winged woodpecker, one of which I observed on a smaller lake
further from the road. My attention was attracted by
its loud note, and I soon perceived the bird clinging
against the extreme summit of a dead pine on the far
side of the lake, at regular intervals giving vent to a
sharp and very powerful call note, repeated several

times in quick succession. It afterwards flew to another tree, settling across a branch instead of on the trunk, and again commencing its cry.

I came across several white-throated sparrows which skulked persistently in the low scrubby growth covering the more open parts of the granite hills, and only announced their presence from time to time by a feeble chirp, very seldom allowing one to obtain a view of them. While wandering over this rocky ground my attention was arrested by a continual tapping at a little distance which I judged to be occasioned by a wood-pecker at first, but soon found that it proceeded from a little dead stump near at hand, and on making my way to it out flew two chickadees one after the other from a small aperture, there being a little cavity within the stump which the birds had been busy excavating. The habit of burrowing in this species is somewhat interesting, as it shows a close affinity in nesting habits to our burrow-ing marsh tit, although the bird itself appears to me to be intermediate between our great tit and coal tit.

Another bird that I noticed here was the yellow-rumped warbler, which drew my attention to itself by its almost aggressive chant, uttered from the upper branches of a little tree. This song, which was rather loud and sharp, might be syllabled as *chi-chi*, *chi-chi*, *chi*, reiterated with almost vehement persistence. It is a very trim and active little bird, and not at all shy if unalarmed.

In some of the more wooded tracts the little black-throated yellow warbler is common about this time. It is one of the prettiest of the North American " warblers," having the upper plumage of a yellowish-green hue, with a very conspicuous lemon-yellow face and a velvety-black throat and upper breast, with also bold black streaks on the flanks. Added to its neat form and tasteful colours it also possesses a pleasing little song, but its ordinary note is a short chirp.

The " white-bill," or black snowbird, breeds com-monly in Nova Scotia, but it was not until May 21 that I succeeded in discovering a nest among the forest-covered hills which are always its favourite resort. This nest was in a ragged bank on the roadside, and was placed upon the upper surface of a projecting piece of rock, being well concealed from view by the overhanging turf above. It had a compact exterior wall of grass stalks and fine roots, except at such portion of its circumference as abutted against the bank behind, while within was a neat and substantial cup of dry grass with a few white horsehairs. I disturbed the female from the nest, which contained four eggs, in appearance of a faint greenish-white, sparingly marked with small specks of reddish-brown and purplish, and with a ring of spots of the same around the larger end.

One of these mornings going down to the North-West Arm I noticed a tree-swallow busily preparing its nest in a tree by the roadside, the site selected being a small hole not more than eight feet from the ground. The mate was meanwhile sitting quietly upon a telegraph wire across the road, quite unalarmed, although apparently eyeing me narrowly from its perch as if in some doubt as to my intentions.

On the road here also I observed one of America's most typical and familiar birds. This was the "oven-bird," a species which has much of the appearance of an *Accentor*, save that its rich golden crown readily distinguishes it. This individual alighted in the road near to me, uttering its loud ascending trill from the ground, and when disturbed it merely flew up and settled a little further along the road, recommencing its remarkable chant.

In the wet woods by Chocolate Lake was a wren (*Troglodytes hiemalis*), very much like our common English species and having a remarkably similar song ; also several red-eyed flycatchers, or vireos, birds of curious appearance which frequented the upper branches of trees, uttering from the sheltering foliage a singular song, consisting of a few liquid notes incessantly repeated ; but it was hard to recognise in its song the supposed resemblance to the *whip-tom-kelly* of the familiar old story.

Happening one day, this same week, to spend an hour or two on board a vessel moored alongside a quay in the harbour, I amused myself for some time by watching the marine life in the still, clear water around the vessel. The whole surface of the water was teeming with a small animal, appertaining to *Medusa*, which consisted of an almost transparent whitish disc, less than an inch in diameter, with two retractile filaments which could be extended to the length of five or six inches, and by the alternate extension and retraction of which the animal moved through the water. I also noticed several small whitish *Medusæ*, and one pink one; these, I observed, moved almost vertically (instead of horizontally) and by regular expansions and contractions of the disc. The wooden piles of the wharf indicated exactly the tidal rise and fall at this part of the world; it was then low water, and I assumed the fall to be about five feet—very different to the tremendous tidal wave which sweeps up the Bay of Fundy, less than a hundred miles away.

On these wooden piles, and above low-water mark, were immense concretions of mussels, but of small size. Below low-water mark the seaweeds grew in profusion, and many forms of marine life could also be seen. Anemones were abundant and of varying sizes, but all were of whitish and brownish hues. Here and there was a starfish, of the forms most common on our own coasts, and deeper down there seemed to be corals or sponges; but there the vision failed, for the light died upon the borders of the depth below and I could see no more.

It was while watching the marvels of this little underworld—so strangely quiet and secluded and so undisturbed by the busy turmoil on the wharf above—that I perceived a long twisting and twining tentacle floating, or rather creeping outward and upward from the forest of seaweed at the side of one of the piles. It was followed by another, and then I discerned the slimy, flabby, whitish body of an octopus appear, but only for a moment, as it soon slowly descended again. The

visible tentacles of this individual must have been very
nearly two feet in length, so that it was by no means a
small one of its kind, and yet far different from the
fearful creatures, born of the imagination rather than
the reality, which are even now believed in, although
still receding from the light of a scientific age. Yet
may not such monsters still be lurking all unknown in
the gloomy depths of that inexhaustible treasure-house
of Nature, which is for ever yielding up its secrets bit
by bit, yet all reluctantly, to the curiosity of inquiring
man ?

CHAPTER XIII.

MUST by no means omit to mention one little hawk, which is to be found breeding not so very uncommonly in some parts of Nova Scotia during the spring. This is the sharp-shinned hawk of Wilson — a dashing and interesting little species, and one, moreover, which appears to have less traducers than others of its kindred. There seems, indeed, little in its habits or mode of living, to gain for it human enemies, its prey comprising for the most part small birds, lizards, red squirrels, or the larger kinds of insects. Its flight is more than usually rapid and erratic; its whole mode of action, indeed, being spirited and fierce.

As with other birds of prey, the female bird is the larger in size, being about thirteen inches in total length, while the male is nearly two inches less, and also differs somewhat in plumage.

I have not myself found the nest of the sharp-shinned hawk in Nova Scotia. The bird is described as preferring a cedar swamp for the purpose of nidification, but pine trees are also undoubtedly made use of. The nest is placed at a height varying from ten or fifteen, up to forty feet or more from the ground. It resembles to some extent the nest of our sparrow-hawk (*Accipiter nisus*), being constructed of sticks, sometimes as much as half an inch in diameter, and neatly lined with small twigs, yet with no softer material than these for the eggs to rest upon. These latter are laid during the latter part of May in the United States, but perhaps a week later in Nova Scotia. They are three, four, or five in number, creamy or greyish white as to ground tint, blotched and streaked with purplish-brown, cinnamon and greyish, the average size of an egg being 1·48 by 1·19 inches. The bird, it is said, almost invariably manifests great indignation when the nest is approached, giving vent to loud cries, and yet always keeping at a sufficiently safe distance from the intruder.

Wilson gave this species the somewhat curious name it bears, on account of its having the "edges of the inside of the shins, below the knee, projecting like the edge of a knife, hard and sharp." The tarsus is, of course, intended, and the term "knee" refers to the tarsal joint or *ankle*. It is strange, it may be remarked, that such a careful and painstaking ornithologist as Alexander Wilson should not have been free from the popular ignorance which still discovers the tibia of a bird in its tarsus, and fails to discern the true "knee," although so very apparent to anyone who will take a dead bird in his hand and make a very cursory examination of its external characteristics.

This lack of acquaintance with anatomy is most lamentable. Few seem to realise to what an extent it can dwarf and confine the mind. There are always to be met many good and capable observers of a grade far removed from the vulgar, yet whose mental vision and reasoning powers are all awry through the want of just a slight study of comparative anatomy. No man or woman can indeed look upon the productions of Nature with any degree of rational understanding unless he or she possesses this slight acquaintance with comparative anatomy, which alone permits the observer to see things intellectually, instead of seeing them with a mere animal vision.

CHAPTER XIV.

N May 28 I started by rail from Halifax to Rimouski, a distance of about 385 miles. The route lay northward through Nova Scotia, through Northern New Brunswick, and also through a part of Quebec, my destination being situated upon the south shore of the Gulf of St. Lawrence, and being a point of call for the homeward-bound liners from Montreal and Quebec.

The scenery for a great part of the distance consisted of alternating forest, lake and river, with occasional ranges of wooded hills, and here and there cabins or fields, and sometimes a considerable settlement. The woods were chiefly coniferous, with occasional tracts of birch or other deodiferous trees. Lumbering was progressing in some parts, and once we passed a small river, which was crowded with logs as far as the view extended. The birds most commonly to be observed were crows and American robins. Once I saw a large blue heron reposing quietly among the straggling reeds at the edge of a lake, and frequently small birds of various species fled on either side into the forest. Horrid, swampy woods were abundant, and here the trees rose from several feet of pestiferous-looking water, filled in every direction with limbs and trunks of trees in every stage of decay, while round about the swamp undergrowth raised itself from the water.

In some parts of Nova Scotia and in the adjoining portion of New Brunswick, are very rich tracts of land formed by the alluvial deposit of the streams, and apparently at one time consisting of mud-flats more or less under water. This formation extends over large tracts of land around the head of the Bay of Fundy, and also around the Petitcodiac, on which stands the town of Moncton. Indeed around the great Basin of Minas, are miles upon miles of mud-flats which would be dry land if it were not for the mighty tidal wave of the Bay of Fundy.

A little before daybreak the next morning we began to enter the rocky region bordering the southern part of the great desolate peninsula of Gaspé, situated in the extreme east of the province of Quebec, and upon the southern side of the Gulf of St. Lawrence. Once we were running for many miles alongside a rapid river, upon the other side of which was a continuous range of the heavily-wooded hills of Gaspé, this stream forming a natural boundary to the vast wilderness beyond.

The morning was well advanced before the train slowed down, and we alighted, stiff and weary, on the station platform in the picturesque little French town which was to mark the first stage of my journey. Rimouski is thoroughly French, there being hardly a resident of British extraction in the place. The town boasts a handsome Roman Catholic church and other buildings, including a large boys' school or college, under the management of the priesthood. I noticed, with pleasure, that the youths had plenty of outdoor exercise and sport together with military drill, and that in spite of the fact that to-day was Sunday. In an enclosure adjoining an inn, also, several persons were enjoying a game of croquet, while many onlookers stood around. Indeed, the general brightness, cleanliness and prosperity of this little town was apparent on every hand.

The beaches left bare by the receding tide were covered with flocks of shore-birds, amongst which I identified some parties of the white-rumped sandpiper and the red-backed sandpiper, or American dunlin, but

5

I could not make out the species of the majority of the
flocks. I also visited the woods in the vicinity of the
town but did not observe much of interest.

We left Rimouski in the small hours of the following
morning on board the s.s. "Circassian," and during
the whole of that day we were ploughing the waters of
the mighty Gulf of St. Lawrence.

The morning of the last day in May broke to find us
still coasting the length of the vast island of Anticosti—
an unvarying range of hills of no great height, yet with
the snow still lying along their summits, to prove the
inhospitable and desolate nature of the island. By
and by all land faded away, and the gulf widened and
stretched away in a great inland sea reaching northward
to Newfoundland and the coast of Labrador, and southward to Cape Breton and the shores of Nova Scotia.
During the day I observed a great number of little auks,
mostly in small parties, although now and again flights
of upwards of a hundred of these curious little birds rose
from the water and flew round the vessel. The only
other birds which came near were a solitary pair of
Brunnich's guillemots : a more frequent species here
than the rather smaller common guillemot.

As evening came on the rays of the setting sun
glittered across the unruffled surface of the water and
played in a halo of glory around the rocky heads of St.
Pierre and Miquelon lying a few miles away to the
northward. Between glistened the white sails of one or
two fishing boats or small coasting vessels. These two
little islets, which dwindled momentarily and passed
gradually into the ruddy glow of the fairyland astern,
possess a peculiar interest. They are all that remain of
the once great North American possessions of France.

We passed Cape Race at five a.m. on June 1 and
were proceeding up the eastern coast of Newfoundland
all the morning. Here we passed a great number of
icebergs, as many as twenty being in sight at once ;
nearly all of them, however, had been driven in-shore,
the coast being dotted with them all along, but some
few were floating down southward some miles from the

coast, and these necessitated constant attention on the part of those in charge of the vessel. The bergs were of every imaginable form and size. One was roughly in the form of a pyramid with the apex broken off, the height from the water-line being nearly three times as much as the breadth.

Most of the larger icebergs evidently consisted of masses detached from the solid ice-field, and these floated in a variety of inclinations; one or two rode in their former horizontal position, but the majority were inclined to one side or another, and some so much so that they presented roughly a " house-top " shape, the original surface constituting one slope and one of the fractured sides the other. On some of the ledges of these masses the sea-birds could be seen congregated in hundreds upon hundreds. The largest berg of all, however, was seen shortly before we entered the harbour of St. John's. It was an immense block, roughly oblong in shape, from the mass of an ice-field, and was floating nearly in its original position, its surface being worn by furrows and corrugations, and much soiled from some cause. The surface of the mass was estimated to be at least an acre in extent; it projected a considerable height from the water, and as all icebergs have two-thirds of their mass below the surface, and only one-third above, the weight of this huge block must have been truly prodigious. The beautifully pinnacled and turreted icebergs common further southward are rare so far north as this, where the sun and ocean lack the warmth necessary to fashion such picturesque and beautiful architecture from these rough-hewn unwieldy blocks.

Among the birds frequenting the coast I noticed several great black-backed gulls, American herring-gulls and kittiwakes, and also a solitary Iceland gull (one of the white-winged group) just on entering the harbour of St. John's. About the fishing-grounds were many greater shearwaters, which were usually to be observed in parties, sometimes as many as fifty in number; they frequently settled upon the surface of the water and sometimes dived.

We had a favourable passage from St. John's, but up to the time we neared the Irish coast very little of interest is to be recorded. In the mid-Atlantic the only birds I noticed were some few of the small, black, white-rumped petrels, including the Leach's, Wilson's, and stormy petrels, which are commonly known as "Mother Cary's chickens," and a small number of greater shearwaters, with which were, I believe, one or two of the less known "dark-bodied" shearwaters, which were until recently thought to be the young of the last mentioned species. There were also a fair number of fulmars, among which were several of the dark phase, which variety or race has those parts which are normally white of a deepish grey.

On June 7 we were off the north-west coast of Ireland, and bird-life was more abundant, although not of great variety. The fulmars were still with us in some numbers, but there were none of the dark phase which I had noticed some few days before. There were also here a few black-backed, herring and other gulls, one pair of gannets, and numbers of puffins, guillemots and other diving birds which were usually observed in parties of varying magnitude.

As we entered the approach to Lough Foyle we were much struck by the beautiful and picturesque scenery, superior, I thought, to almost anything to be seen on the other side of the water. The hillsides were green with the greenest of well-tended fields, and above these the hills, green to the very summit, seemed to assert the peculiar aptness of the epithet "Emerald Isle." Through the fields wound white and tiresome-looking roads, while here and there were dotted the characteristic cabins of the peasantry, who were summoned to their doors by the report of our rockets and stood there waving handkerchiefs and aprons to welcome home again countrymen, or, perhaps, relations. Down by the water were the residences of the more wealthy classes, and here, too, we passed a small but picturesque ruin which instantly became an object of interest to our amateur photographer. Soon, however, we were lying-to in Lough Foyle itself, which extended before us for miles, its glassy surface glittering like burnished silver in the rays of the declining sun, and its unruffled expanse broken only by one or two motionless small craft. After a far too limited stay at this interesting spot, however, we were once more on our way to old England.

We passed the Manxmen's shores at an early hour the next morning, and before 8 a.m. were steaming up the Mersey in the midst of a characteristically dense fog, which, however, lifted sufficiently for us to catch a glimpse of New Brighton as we passed; while in a few minutes more we were lying-to off the docks at Liverpool. Then came the disembarking, the ordeal of the custom-house, and the bustle of the railway station, breaking up the glamour of solitude as a dream dies at daybreak, and speeding us back along one of those great veins to be re-energised in this great throbbing heart which is for ever pulsing forth its arteries into the most distant regions of the earth!

APPENDIX.

The following is a systematic list of the species of North American birds mentioned in the foregoing pages, the prefixed numbers and the nomenclature being in accordance with the check list issued by the American Ornithologists' Union, and the square-bracketed synonyms being those more usually employed by British ornithologists. Round brackets indicate that the name is not that given by the original describer. The list is not intended as a complete list of Nova Scotian birds, there being many species occurring in the province which have not been mentioned in the present work.

Marked * are species which also occur in Great Britain.
Marked † are racial varieties of species which occur in Britain.

6. **Podilymbus podiceps** (Linn.), Pied-billed Grebe.

7. ****Urinator imber** (Gunn), Loon. [*Colymbus glacialis*, Linn., Great Northern Diver.]

11. ****Urinator lumme** (Gunn), Red-throated Loon. [*Colymbus septentrionalis*, Linn., Red-throated Diver.]

13. ****Fratercula arctica** (Linn.), Puffin.

27. ****Cepphus grylle** (Linn.), Black Guillemot. [*Uria grylle.*]

30. ****Uria troile** (Linn.), Murre. [Common Guillemot.]

31. ****Uria lomvia** (Linn.), Brunnich's Murre. [*Uria bruennichi*, E. Sabine, Brunnich's Guillemot.]

34. *Alle alle (Linn.), DOVEKIE. [*Mergulus alle* (Linn.), Little Auk.]
38. *Stercorarius longicaudus, Vieill., LONG-TAILED JAEGER. [*Stercorarius parasiticus* (Linn.), Long-tailed Skua ; Buffon's Skua. Note.—The *S. parasiticus* of the A.O.U. list is the Arctic or Richardson's Skua, *S. crepidatus* of British ornithologists.]
40. *Rissa tridactyla (Linn.), KITTIWAKE.
43. *Larus leucopterus, Faber, ICELAND GULL.
47. *Larus marinus, Linn., GREAT BLACK-BACKED GULL.
51*a*.†Larus argentatus smithsonianus, Coues, AMERICAN HERRING GULL. [*Larus argentatus;* the American form not considered distinct by British ornithologists.]
86. *Fulmarus glacialis (Linn.), FULMAR.
89. *Puffinus major, Faber, GREATER SHEARWATER.
95. *Puffinus griseus (Gmel.), DARK-BODIED SHEARWATER. [Sooty Shearwater.]
104. *Procellaria pelagica, Linn., STORMY PETREL.
106. *Oceanodroma leucorhoa (Vieill.), LEACH'S PETREL. [*Cymochorea leucorrhoa* (Vieill.), Fork-tailed Petrel.]
109. *Oceanites oceanicus (Kuhl), WILSON'S PETREL.
117. *Sula bassana (Linn.), GANNET.
155. *Histrionicus histrionicus (Linn.), HARLEQUIN DUCK. [*Casmonetta histrionica* (Linn.); *Histrionicus minutus* (Linn.)]
172. Branta canadensis (Linn.), CANADA GOOSE. [*Bernicla canadensis* (Linn.).]
194. Ardea herodias, Linn., GREAT BLUE HERON.
240. *Tringa fuscicollis, Vieill., WHITE-RUMPED SANDPIPER. [Bonaparte's Sandpiper.]
243*a*.†Tringa alpina pacifica (Coues), RED-BACKED SANDPIPER. [American Dunlin ; doubtfully distinct from *Tringa alpina.*]
263. Actitis macularia (Linn.), SPOTTED SANDPIPER. [*Totanus macularius* (Linn.)]
298. Dendragapus canadensis (Linn.), CANADA GROUSE. [Spruce Partridge.]

300*a*. **Bonasa umbellus** togata (Linn.), CANADIAN RUFFED GROUSE. [Canadian birds are said to constitute a darker race than the type, *B. umbellus.*]

331. **Circus hudsonius** (Linn.), MARSH HAWK.

332. **Accipiter velox** (Wils.), SHARPSHINNED HAWK.

334. **Accipiter atricapillus** (Wils.), AMERICAN GOS-HAWK. [*Astur atricapillus* (Wils.).]

339. **Buteo lineatus** (Gmel.), RED-SHOULDERED HAWK.

352. **Haliætus leucocephalus** (Linn.), BALD EAGLE.

360. **Falco sparverius**, Linn., AMERICAN SPARROW-HAWK.

364. †**Pandion haliætus carolinensis** (Gmel.), AMERI-CAN OSPREY. [*Pandion haliætus* (Linn.); the separation of the American bird is not recognised by British ornithologists.]

368. **Syrnium nebulosum** (Forst.), BARRED OWL. [*Strix nebulosa* of Forster.]

370. **Ulula cinerea** (Gmel.), GREAT GRAY OWL.

372. **Nyctala acadica** (Gmel.), SAW-WHET OWL.

375. **Bubo virginianus** (Gmel.), GREAT HORNED OWL.

376. *****Nyctea nyctea** (Linn.), SNOWY OWL. [*Nyctea scandiaca* (Linn.).]

390. **Ceryle alcyon** (Linn.), BELTED KINGFISHER.

394. **Dryobates pubescens** (Linn.), DOWNY WOOD-PECKER. [*Picus pubescens* of Linnæus.]

412. **Colaptes auratus** (Linn.), FLICKER. [Golden-winged Woodpecker.]

420. **Chordeiles virginianus** (Gmel.), NIGHT-HAWK.

428. **Trochilus colubris**, Linn., RUBY-THROATED HUM-MING BIRD.

474. *****Otocoris alpestris** (Linn.), HORNED LARK. [Shore Lark.]

477. **Cyanocitta cristata** (Linn.), BLUE JAY.

484. **Perisoreus canadensis** (Linn.), CANADA JAY.

486. †**Corvus corax principalis**, Ridgw., NORTHERN RAVEN. [This race was described since the publication of the A.O.U. list, in which No. 486 is called "American Raven".]

488. **Corvus americanus**, Aud., AMERICAN CROW.

515. †**Pinicola enucleator canadensis** (Cab.), Ameri-can Pine Grosbeak. [*Pyrrhula enucleator* (Linn.); British ornithologists do not recognise the separation of the American bird.]

— *Passer **domesticus** (Linn.), European House Sparrow. [Introduced.]

534. *Plectrophenax **nivalis** (Linn.), Snowflake. [Snow Bunting.]

558. **Zonotrichia albicollis** (Gmel.), White-throated Sparrow.

560. **Spizella socialis** (Wils.), Chipping Sparrow.

567. **Junco hyemalis** (Linn.), Slate-Colored Junco. [Black Snowbird.]

585. **Passerella iliaca** (Merr.), Fox Sparrow.

613. **Chelidon erythrogaster** (Bodd.), Barn Swallow. [*Hirundo horreorum* (Bodd.).]

614. **Tachycincta bicolor** (Vieill.), Tree Swallow. [White-bellied Swallow.]

621. **Lanius borealis** (Vieill.), Northern Shrike.

624. **Vireo olivaceus** (Linn.), Red-eyed Vireo. [Red-eyed Flycatcher.]

636. **Mniotilta varia** (Linn.), Black and White Warbler.

655. **Dendroica coronata** (Linn.), Myrtle Warbler. [Yellow-rumped Warbler.]

667. **Dendroica virens** (Gmel.), Black-throated Green Warbler.

671. **Dendroica vigorsii** (Aud.), Pine Warbler. [*Dendrœca pinus* of Baird; Pine-creeping Warbler.]

674. **Seiurus aurocapillus** (Linn.), Oven-Bird. [Golden crowned Thrush.]

697. **Anthus pensilvanicus** (Lath.), American Pipit. [*Anthus ludovicianus* of Lichtenstein.]

722. **Troglodytes hiemalis**, Vieill., Winter Wren.

726. †**Certhia familiaris americana** (Bonap.), Brown Creeper.

727. **Sitta carolinensis**, Lath., White-breasted Nut-hatch.

728. **Sitta canadensis**, Linn., Red-Breasted Nut-hatch.

735. **Parus atricapillus**, Linn., CHICKADEE. [Black-capped Chickadee.]

740. **Parus hudsonicus**, Forst., HUDSONIAN CHICKADEE.

748 **Regulus satrapa**, Licht., GOLDEN-CROWNED KINGLET.

758a. **Turdus ustulatus swainsonii** (Cab.), OLIVE-BACKED THRUSH.

759b. **Turdus aonalaschkæ pallasii** (Cab.), HERMIT THRUSH.

761. **Merula migratoria** (Linn.), AMERICAN ROBIN. [*Turdus migratorius*, Linn., Migratory Thrush.]

www.ingramcontent.com/pod-product-compliance
Lightning Source LLC
Chambersburg PA
CBHW021955190326
41519CB00009B/1267